Machine Learning
For Financial
Engineering

Advances in Computer Science and Engineering: Texts

Editor-in-Chief: Erol Gelenbe *(Imperial College London)*
Advisory Editors: Manfred Broy *(Technische Universität München)*
Gérard Huet *(INRIA)*

MACHINE LEARNING FOR FINANCIAL ENGINEERING

László Györfi

Budapest University of Technology and Economics, Hungary

György Ottucsák

Budapest University of Technology and Economics, Hungary

Harro Walk

Universität Stuttgart, Germany

Imperial College Press

ICP

Published by

Imperial College Press
57 Shelton Street
Covent Garden
London WC2H 9HE

Distributed by

World Scientific Publishing Co. Pte. Ltd.
5 Toh Tuck Link, Singapore 596224
USA office: 27 Warren Street, Suite 401-402, Hackensack, NJ 07601
UK office: 57 Shelton Street, Covent Garden, London WC2H 9HE

British Library Cataloguing-in-Publication Data
A catalogue record for this book is available from the British Library.

ISBN-13 978-1-84816-813-8
ISBN-10 1-84816-813-6

Printed in Singapore.

Preface

The main purpose of this volume is to investigate algorithmic methods based on machine learning in order to design sequential investment strategies for financial markets. Such sequential investment strategies use information collected from the market's past and determine, at the beginning of a trading period, a portfolio; that is, a way to invest the currently available capital among the assets that are available for purchase or investment.

Our aim in writing this volume is to produce a self-contained text intended for a wide audience, including graduate students in finance, statistics, mathematics, computer science, and engineering, as well as researchers in these fields. Thus, the material is presented in a manner that requires only a basic knowledge of probability.

In the approach we adopt, the goal of the decider or investor is to maximize his wealth in the *long run*. However, the investor does not have direct information about the underlying distributions that are generating the stock prices. In the area of mathematical finance, most of the known theoretical results have been obtained for models that consider single assets in a single period, and they typically assume a parametric model of the underlying stochastic process of the prices.

In the last decade, it has become clear that decision schemes that consider multiple assets simultaneously, and also try to consider decisions over multiple periods, can increase the investor's wealth through judicious rebalancing of investments between the assets. Since accurate statistical modelling of stock market behavior is now known to be notoriously difficult, in our work we take an extreme point of view and work with minimal assumptions on the probabilistic distributions regarding the time series of interest. Our approach addresses the best rebalancing of portfolio assets in the sense that it maximizes the expected log-return. If the distributions of the underlying price processes are unknown, one has to "learn" the optimal portfolio from past data, and effective empirical strategies can then be derived using methods from nonparametric statistical smoothing and machine learning.

The growth-optimal portfolio (GOP) is defined as having a maximal expected growth rate over any time horizon. As a consequence, this portfolio is certain to outperform any other significantly different strategy as the time horizon increases. This property, in particular, has fascinated many researchers in finance and mathematics, and has created an exciting literature on growth optimal investment.

Therefore, Chapter 1 attempts to provide a comprehensive survey of the literature and applications of the GOP. In particular, the heated debate of whether the GOP has a special place among portfolios in the asset allocation decision is reviewed, as this still seems to be an area where some misconceptions exist. The survey also provides a review of the recent use of the GOP as a pricing tool; for example, in the "benchmark approach".

Chapter 2 provides a survey of sequential investment strategies for financial markets. The GOP can be derived from the log-optimal criterion (also called the Kelly-criterion), which means that one chooses the portfolio that maximizes the conditional expectation of the log-return given the past data. Under the memoryless assumption on the underlying market process of the assets' relative prices, the best constantly rebalanced portfolio is studied (the called log-optimal portfolio), which achieves the maximal asymptotic average growth rate. Semi-log optimal portfolio selection as a small computational complexity alternative of the log-optimal portfolio selection is studied both theoretically and empirically. For generalized dynamic portfolio selection, when the market process is stationary and ergodic, the challenging problem is whether or not it is possible to learn the conditional distributions from data, i.e., whether one can construct empirical (data-driven) strategies that achieve the optimal growth rate. It emerges that, by utilizing the current approaches of nonparametric estimates and machine learning algorithms, such empirical GOPs exist. The empirical performance of the methods is illustrated for New-York Stock Exchange (NYSE) data.

The theoretical and empirical optimality of GOP will be based on some assumptions, the most important being that the transaction cost is ignored. In Chapter 3, the discrete time growth optimal investment with proportional transaction costs is considered. Here, the market process is modeled by a first-order Markov process. Assuming that the distribution of the market process is known, we show sequential investment strategies such that, in the long run, the growth rate on trajectories achieves the maximum with probability 1.

In the previous chapters, the model does not include the possibility of short selling and leverage. Chapter 4 revisits the GOP on non-leveraged, long-only markets for memoryless market processes. We derive optimality conditions for frameworks on leverage and short selling, and establish no-ruin conditions. Moreover, we investigate the strategy and its asymptotic growth rate from both theoretical and empirical points of view. The empirical performance of the methods is illustrated for NYSE data showing, that with leverage, the growth rate is drastically increasing.

For constructing empirical GOPs, the role of nonparametric estimates and machine learning algorithms is important. Chapter 5 is devoted to the application of these principles for the prediction of stationary time series. This chapter presents simple procedures for the prediction of a real-valued time series with side information. For the regression problem, we survey the basic principles of nonparametric estimates. Based on current machine learning algorithms, the predictions are the aggregations of several simple predictors. We show that, if the sequence is a realization of a stationary random process, the average of squared errors converges, almost surely, to that of the optimum, given by the Bayes predictor. We offer an analog result for the prediction of Gaussian processes.

Chapter 6 deals with empirical pricing American options, which can be viewed as an optimal stopping problem derived from a backward recursion such that in each step of the recursion one needs conditional expectations. For empirical pricing, Longstaff and Schwartz suggested replacing the conditional expectations by regression function estimates. We survey the current literature and, based on nonparametric regression estimates, some new algorithms are introduced and investigated.

As we conclude this preface, we would like to acknowledge the contribution of many people who influenced this volume. A number of colleagues and friends have, often without realizing it, contributed to our understanding of rebalancing, nonparametrics and machine learning. In particular we would like to thank in this respect Paul Algoet, Andrew Barron, Tom Cover, Miguel Delgado, Luc Devroye, Jürgen Dippon, László Gerencsér, András György, Péter Kevei, Jussi Klemelä, Michael Kohler, Sándor Kolumbán, Adam Krzyżak, Kasper Larsen, Tamás Linder, Gábor Lugosi, Gábor Molnár-Sáska, Eckhard Platen, Miklós Rásonyi, Christian Riis Flor, Walter Schachermayer, Dominik Schäfer, Wolfgang Stummer, Csaba Szepesvári, Frederic Udina, István Vajda and Sara van de Geer.

We would like to thank Erol Gelenbe, the series editor of Imperial College Press, for his initial and never fading encouragment and support throughout this project. Our contact with ICP through Sarah Haynes and Steven Patt has always been pleasant and constructive. Lindsay Robert Wilson's heroic copyeditorial job was greatly appreciated.

In addition, we gratefully acknowledge the research support of the Budapest University of Technology and Economics, the Hungarian Academy of Sciences (MTA SZTAKI), Morgan Stanley, MSCI Inc., the Universität Stuttgart and the University of Southern Denmark. This work is connected to the scientific program of the "Development of quality-oriented and cooperative R+D+I strategy and functional model at BME" project, which is supported by the New Hungary Development Plan (Project ID: TÁMOP-4.2.1/B-09/1/KMR-2010-0002), and supported by the PASCAL2 Network of Excellence under EC grant no. 216886.

László Györfi, Budapest, Hungary
György Ottucsák, Budapest, Hungary
Harro Walk, Stuttgart, Germany

July 17, 2011

Contents

Chapter 1

On the History of the Growth-Optimal Portfolio

Morten Mosegaard Christensen*

Danske Bank A/S
Holmens Kanal 2-12, 1092 København, Denmark.
morten.mosegaard@danskebank.dk

The growth-optimal portfolio (GOP) is a portfolio which has a maximal expected growth rate over any time horizon. As a consequence, this portfolio is certain to outperform any other significantly different strategy as the time horizon increases. This property, in particular, has fascinated many researchers in finance and mathematics and has created a huge and exciting literature on growth-optimal investment. This chapter attempts to provide a comprehensive survey of the literature and applications of the GOP. In particular, the heated debate of whether the GOP has a special place among portfolios in the asset allocation decision is reviewed, as this still seems to be an area where some misconceptions exist. The survey also provides an extensive review of the recent use of the GOP as a pricing tool, in, for example, the so-called "benchmark approach". This approach builds on the numéraire property of the GOP, that is, the fact that any other asset denominated in units of the GOP becomes a supermartingale.

1.1. Introduction and Historical Overview

Over the past 50 years, a large number of papers have investigated the GOP. As the name implies, this portfolio can be used by an investor to maximize the expected growth rate of his portfolio. However, this is only one of many uses for this object. In the literature, it has been applied in such diverse areas as portfolio theory, gambling, utility theory, information theory, game theory, theoretical and applied asset pricing, insurance, capital structure theory, macro-economy and event studies. The ambition of the

---

*The views in this paper represent the views of the author alone and do not represent the views of Danske Bank A/S or any of its affiliates.

present chapter is to present a reasonably comprehensive review of the different connections in which the portfolio has been applied. An earlier survey in [Hakansson and Ziemba (1995)] focused mainly on the applications of the GOP for investment and gambling purposes. Although this will be discussed in Section 1.3, the present chapter has a somewhat wider scope.

The origins of the GOP have usually been tracked to the paper [Kelly (1956)], hence the name "Kelly criterion", which is used synonymously. (The name "Kelly criterion" probably originates from [Thorp (1971)].) Kelly's motivation came from gambling and information theory, and his paper derived a striking but simple result: there is an optimal gambling strategy, such that, with probability one, this optimal gambling strategy will accumulate more wealth than any other different strategy. Kelly's strategy was the GOP and in this respect the GOP was discovered by him. However, whether this is the true origin of the GOP depends on a point of view. The GOP is a portfolio with several aspects, one of which is the maximization of the *geometric mean*. In this respect, the history might be said to have its origin in [Williams (1936)], who considered speculators in a multi-period setting and reached the conclusion that, due to compounding, speculators should consider the geometric and not the arithmetic mean. Williams did not reach any result regarding the growth properties of this approach but was often cited as the earliest paper on the GOP in the seventies seemingly due to the remarks on the geometric mean made in the appendix of his paper. Yet another way of approaching the history of the GOP is from the perspective of utility theory. As the GOP is the choice of a log-utility investor, one might investigate the origin of this utility function. In this sense, the history dates even further back to the 18th century. The mathematician Pierre Rèmond Montemort challenged Nicolas Bernoulli with five problems, one of which was the famous St. Petersburg paradox. The St. Petersburg paradox refers to the coin tossing game, where returns are given as 2^{n-1}, where n is the number of games before "heads" come up the first time. The expected value of participating is infinite, but in Nicolas Bernoulli's words, no sensible man would pay 20 dollars for participating. Nicolas Bernoulli posed the problem to his cousin, Daniel Bernoulli, who suggested using a utility function to ensure that (rational) gamblers would use a more conservative strategy. Note that any unbounded utility function is subject to the generalized St. Petersburg paradox, obtained by scaling the outcomes of the original paradox sufficiently to provide infinite expected utility. For more information see, e.g., [Bernoulli (1954)], [Menger (1967)], [Samuelson (1977)] or [Aase (2001)]. Nicolas Bernoulli conjectured that gamblers should be risk averse, but less so if they had high wealth. In particular, he suggested that marginal utility should be inversely proportional to wealth, which is tantamount to assuming log-utility. However, the choice of logarithm appears to have nothing to do with the growth properties of this strategy, as is sometimes suggested. The original article "Specimen Theoriae Nova de Mensura Sortis" from 1738 is reprinted in Econometrica [Bernoulli (1954)] and do not mention growth. (For information about recent developments on St. Petersburg portfolio games, see Chapter 2 of this volume.) It should be noted that the St. Petersburg paradox was resolved even earlier by Cramer, who used the square-root function in a similar way. Hence log-utility has a history going back at least 250 years and, in this sense, so has the GOP. It seems to have been Bernoulli who, to some extent, inspired the article [Latané (1959)]. Independent of Kelly's result, Latané suggested that investors should maximize the geometric mean of their portfolios, as this would maximize the probability that the portfolio would be more valuable than any other portfolio. [Latané (1959)] has a reference to Kelly's 1956 paper, but Latané mentions that he was unaware of Kelly's result before presenting the paper at an earlier conference in 1956. Regardless of where the history of the GOP is said to start, real interest in the GOP did not arise until after the papers by Kelly and Latané. As will be described later on, the goal suggested by Latané caused a great deal of debate among economists, which has not completely died out yet. The paper by Kelly caused a great deal of immediate interest in the mathematics and gambling community. [Breiman (1960, 1961)] expanded the analysis of [Kelly (1956)] and discussed applications for long-term investment and gambling in a more general mathematical setting.

Calculating the GOP is generally very difficult in discrete time and is treated in [Bellman and Kalaba (1957)], [Elton and Gruber (1974)] and [Maier et al. (1977b)], although the difficulties disappear whenever the market is complete. This is similar to the case when jumps in asset prices happen at random. In the continuous-time continuous-diffusion case, the problem is much easier and was solved in [Merton (1969)]. This problem, along with a general study of the properties of the GOP, has been studied for decades and is still being studied today. Mathematicians fascinated by the properties of the GOP have contributed to the literature a significant number of theoretical articles spelling out the properties of the GOP in a variety of scenarios and increasingly generalized settings, including continuous-time models based on semimartingale representation of asset prices. Today, solutions to the problem exist in a semi-explicit form and, in the general case, the GOP can be characterized in terms of the semimartingale characteristic triplet. A non-linear integral equation must still be solved to obtain the portfolio weights. The properties of the GOP and the formulas required to calculate the strategy in a given situation are discussed in Section 1.2, which has been split into two parts. Section 1.2.1 deals with the simple discrete-time case, providing the main properties of the GOP without the need for demanding mathematical techniques. Section 1.2.2 deals with the fully-general case, where asset-price processes are modeled as semimartingales, and contains examples of important special cases.

The growth optimality and the properties highlighted in Section 1.2 inspired authors to recommend the GOP as a universally "best" strategy and this sparked a heated debate. In a number of papers, Paul Samuelson and other academics argued that the GOP was only one among many other investment rules and any belief that the GOP was universally superior rested on a fallacy; see, for example, [Samuelson (1963)]. The substance of this discussion is explained in detail in Section 1.3.1. The debate from the late sixties and seventies contains some important lessons to be remembered when discussing the application of the GOP as a long-term investment strategy.

The use of the GOP became referred to as the *growth-optimum theory* and it was introduced as an alternative to expected utility and the mean-variance approaches to asset pricing. It was argued that a theory for portfolio selection and asset pricing based on the GOP would have properties which are more appealing than those implied by the mean-variance approach developed by [Markowitz (1952)]. Consequently, a significant amount of the literature deals with a comparison of the two approaches. A discussion of the relation between the GOP and the mean-variance model is presented in Section 1.3.2. Since a main argument for applying the GOP is its ability to outperform other portfolios over time, authors have tried to estimate the time needed to be "reasonably" sure to obtain a better result using the GOP. Some answers to this question are provided in Section 1.3.3.

The fact that asset prices, when denominated in terms of the GOP, become supermartingales, was realized quite early, appearing in a proof in [Breiman (1960)][Theorem 1]. It was not until 1990 in [Long (1990)] that this property was given a more thorough treatment. Although Long suggested this as a method for measuring abnormal returns in event studies and this approach has been followed recently in working papers by [Gerard *et al.*

(2000)] and [Hentschel and Long (2004)], the consequences of the numéraire property stretch much further. It suggested a change of numéraire technique for asset pricing under which a change of probability measure would be unnecessary. The first time this is treated explicitly appears to be in [Bajeux-Besnaino and Portait (1997a)] in the late nineties. At first, the use of the GOP for derivative pricing purposes was essentially just the choice of a particular pricing operator in an incomplete market. Over the past five years, this idea became developed further in the benchmark framework of [Platen (2002)] and later papers, who emphasize the applicability of this idea in the absence of a risk-neutral probability measure. The use of the GOP as a tool for derivative pricing is reviewed in Section 1.4. This has motivated a substantial part of this chapter, because it essentially challenges the approach of using some risk-neutral measure for pricing derivatives. During this chapter we are going to conduct a hopefully thorough analysis of what arbitrage concepts are relevant in a mathematically-consistent theory of derivative pricing and what role martingale measures play in this context. Section 1.4 gives a motivation and foreshadows some of the results derived later on. A complete survey of the benchmark approach is beyond the scope of this chapter, but may be found in [Platen (2006a)].

The suggestion that such GOP-denominated prices could be martingales is important to the empirical work, since this provides a testable assumption which can be verified from market data. The Kuhn–Tucker conditions for optimality provide only the supermartingale property which may be a problem; see Section 1.2 and Section 1.5. Few empirical papers exist, and most appeared during the seventies. Some papers tried to obtain evidence for or against the assumption that the market was dominated by growth optimizers and to see how the growth-optimum model compared to the mean-variance approach. Others tried to document the performance of the GOP as an investment strategy, in comparison with other strategies. Section 1.5 deals with the existing empirical evidence related to the GOP.

Since an understanding of the properties of the GOP provides a useful background for analyzing its applications, the first task will be to present the relevant results which describe some of the remarkable properties of the GOP. The next section is separated into a survey of discrete-time results which are reasonably accessible and a more mathematically-demanding survey in continuous time. This is not just mathematically convenient, but also fairly chronological. It also discusses the issues related to *solving* for the GOP strategy, which is a non-trivial task in the general case. Readers that are particularly interested in the GOP from an investment perspective may prefer to skip the general treatment in Section 1.2.2 with very little loss. However, most of this chapter relies extensively on the continuous-time analysis and later sections build on the results obtained in Section 1.2.2. Extensive references will be given in the notes at the end of each section and only the most important references are kept within the main text, in order to keep it fluent and short.

1.2. Theoretical Studies of the GOP

The early literature on the GOP was usually framed in discrete time and considered a restricted number of distributions. Despite the simplicity and loss of generality, most of the interesting properties of the GOP can be analyzed within such a framework. The more recent theory has almost exclusively considered the GOP in continuous time and considers very general cases, requiring the machinery of stochastic integration and sometimes applies a very general class of processes, semimartingales, which are well-suited to financial modelling. Although many of the fundamental properties of the GOP carry over to the general case, there are some quite technical, but very important differences from the discrete-time case.

Section 1.2.1 reviews the major theoretical properties of the GOP in a discrete-time framework, requiring only basic probability theory. Section 1.2.2, on the other hand, surveys the GOP problem in a very general semimartingale setting and places modern studies within this framework. It uses the theory of stochastic integration with respect to semimartingales, but simpler examples have been provided for illustrative purposes. Both sections are structured around three basic issues:

- *Existence*, which is fundamental, particularly for theoretical applications.
- *Growth properties* are those that are exploited when using the GOP as an investment strategy.
- Finally, the *numéraire property* which is essential for the use of the GOP in derivative pricing.

1.2.1. *Discrete Time*

Consider a market consisting of a finite number of non-dividend paying assets. The market consists of $d+1$ assets, represented by a $d+1$ dimensional vector process, S, where

$$S = \left\{ S(t) = (S^{(0)}(t), \ldots, S^{(d)}(t)), \ t \in \{0, 1, \ldots, T\} \right\}. \qquad (1.1)$$

The first asset, $S^{(0)}$, is sometimes assumed to be risk-free from one period to the next, i.e., the value $S^{(0)}(t)$ is known at time $t-1$. In other words, $S^{(0)}$ is a predictable process. Mathematically, let $(\Omega, \mathcal{F}, \underline{\mathcal{F}}, \mathbb{P})$ denote a filtered probability space, where $\underline{\mathcal{F}} = (\mathcal{F}_t)_{t \in \{0,1,\ldots,T\}}$ is an increasing sequence of information sets. Each price process $S^{(i)} = \{S^{(i)}(t), \ t \in \{0, 1, \ldots, T\}\}$ is assumed to be adapted to the filtration $\underline{\mathcal{F}}$. In other words, the price of each asset is known at time t, given the information $\mathcal{F}_t$. Sometimes it will be convenient to work on an infinite time horizon, in which case $T = \infty$. However, unless otherwise noted, T is assumed to be some finite number.

Define the *return* process

$$R = \left\{ R(t) = (R^0(t), \ldots, R^d(t)), \ t \in \{1, 2, \ldots, T\} \right\}$$

by $R^i(t) \triangleq \frac{S^{(i)}(t)}{S^{(i)}(t-1)} - 1$. Often, it is assumed that returns are independent over time, and, for simplicity, this assumption is made in this section.

Investors in such a market consider the choice of a *strategy*

$$\delta = \left\{ \delta(t) = (\delta^{(0)}(t), \ldots, \delta^{(d)}(t)), \ t \in \{0, \ldots, T\} \right\},$$

where $\delta^{(i)}(t)$ denotes the number of units of asset i that are being held during the period $(t, t+1]$. As usual, some notion of "reasonable" strategy has to be used. Definition 1.1 makes this precise.

Definition 1.1. A trading strategy, δ, generates the portfolio value process $S^{(\delta)}(t) \triangleq \delta(t) \cdot S(t)^\mathsf{T}$, where $\cdot$ denotes the standard Euclidean inner product. The strategy is called *admissible* if it satisfies the three conditions below.

    (1) Non-anticipative: The process δ is adapted to the filtration $\mathcal{F}$, meaning that $\delta(t)$ can only be chosen based on information available at time t.

    (2) Limited liability: The strategy generates a portfolio process $S^{(\delta)}(t)$ which is nonnegative.

    (3) Self-financing: $\delta(t-1) \cdot S(t) = \delta(t) \cdot S(t)$, $t \in \{1, \ldots T\}$ or equivalently $\Delta S^{(\delta)}(t) = \delta(t-1) \cdot \Delta S(t)$.

The set of admissible portfolios in the market will be denoted $\Theta(S)$, and $\underline{\Theta}(S)$ will denote the strictly positive portfolios. It is assumed that $\underline{\Theta}(S) \neq \emptyset$, where $\emptyset$ is the empty set.

These assumptions are fairly standard. The first part assumes that any investor is unable to look into the future, i.e. only the current and past information is available. The second part requires the investor to remain solvent, since his total wealth must always be nonnegative. This requirement will prevent him from taking an unreasonably risky position. Technically, this constraint is not strictly necessary in the very simple case described in this subsection, unless the time horizon T is infinite. The third part requires that the investor re-invests all money in each time step. No wealth is withdrawn or added to the portfolio. This means that intermediate consumption is not possible. Although this is a restriction in generality, consumption can be allowed at the cost of slightly more complex statements. Since consumption is not important for the purpose of this chapter, we have decided to leave it out altogether. The requirement that it should be possible to form a strictly positive portfolio is important, since the growth rate of any portfolio with a chance of defaulting will be minus infinity.

Consider an investor who invests a dollar of wealth in some portfolio. At the end of period T, his wealth becomes

$$S^{(\delta)}(T) = S^{(\delta)}(0) \prod_{i=1}^{T}(1 + R^{(\delta)}(i)),$$

where $R^{(\delta)}(t)$ is the return in period t. If the portfolio *fractions* are fixed during the period, the right hand side is the product of T independent and identically distributed (i.i.d.) random variables. The *geometric average* return over the period is then

$$\left(\prod_{i=1}^{T} \left(1 + R^{(\delta)}(i) \right) \right)^{\frac{1}{T}}.$$

Because the returns of each period are i.i.d., this average is a sample of the *geometric mean value* of the one-period return distribution. For discrete random variables, the geometric mean of a random variable X taking (not necessarily distinct) values $x_1, \ldots x_S$, with equal probabilities, is defined as

$$G(X) \triangleq \left(\Pi_{s=1}^{S} x_s \right)^{\frac{1}{S}} = \left(\Pi_{k=1}^{K} \tilde{x}_k^{f_k} \right) = \exp(\mathbb{E}[\log(X)]),$$

where $\tilde{x}_k$ are the distinct values of X and f_k is the frequency at which $X = x_k$, that is $f_k = \mathbb{P}(X = x_k)$. In other words, the geometric mean is the exponential function of the *growth rate* $g^{\delta}(t) \triangleq \mathbb{E}[\log(1 + R^{(\delta)}(t))]$ of some portfolio. Hence, if Ω is discrete, or, more precisely, if the σ-algebra $\mathcal{F}$ on Ω

is countable, maximizing the geometric mean is equivalent to maximizing the expected growth rate. Generally, one defines the geometric mean of an arbitrary random variable by

$$G(X) \triangleq \exp(\mathbb{E}[\log(X)]),$$

assuming the mean value $\mathbb{E}[\log(X)]$ is well-defined. Over long stretches, intuition dictates that each realized value of the return distribution should appear, on average, the number of times dictated by its frequency, and hence, as the number of periods increase, it would hold that

$$\left(\prod_{i=1}^{T} (1 + R^{(\delta)}(i)) \right)^{\frac{1}{T}} = \exp \left(\frac{\sum_{i=1}^{T} \log(1 + R^{(\delta)}(i))}{T} \right) \to G(1 + R^{(\delta)}(1))$$

as $T \to \infty$. This states that the average growth rate converges to the expected growth rate. In fact, this heuristic argument can be made precise by an application of the law of large numbers, but here we only need it for establishing intuition. In multi-period models, the geometric mean was suggested by [Williams (1936)] as a natural performance measure, because it took into account the effects from compounding. Instead of worrying about the average expected return, an investor who invests repeatedly should worry about the geometric mean return. As it will be discussed later on, not everyone agreed with this idea, but it explains why one might consider the problem

$$\sup_{S^{(\delta)}(T) \in \Theta} \mathbb{E} \left[\log \left(\frac{S^{(\delta)}(T)}{S^{(\delta)}(0)} \right) \right]. \tag{1.2}$$

Definition 1.2. A solution, $S^{(\hat{\delta})}$, to (1.2) is called a GOP.

Hence, the objective given by (1.2) is often referred to as the *geometric mean criteria*. Economists may view this as the maximization of expected terminal wealth for an individual with logarithmic utility. However, it is important to realize that the GOP was introduced into economic theory, not as a special case of a general utility maximization problem, but because it seems an intuitive objective when the investment horizon stretches over several periods. The next section will demonstrate the importance of this observation. For simplicity, it is always assumed that $S^{(\delta)}(0) = 1$, i.e. the investors start with one unit of wealth.

If an investor can find an admissible portfolio having zero initial cost and which provides a strictly positive pay-off at some future date, a solution to (1.2) will not exist. Such a portfolio is called an *arbitrage* and is formally defined in the following way.

Definition 1.3. An admissible strategy δ is called an arbitrage strategy if

$$S^{(\delta)}(0) = 0 \quad \mathbb{P}(S^{(\delta)}(T) \geq 0) = 1 \quad \mathbb{P}(S^{(\delta)}(T) > 0) > 0.$$

It seems reasonable that this is closely related to the existence of a solution to problem (1.2), because the existence of a strategy that creates "something out of nothing" would provide an infinitely high growth rate. In fact, in the present discrete time case, the two definitions are completely equivalent.

Theorem 1.1. *There exists a GOP, $S^{(\underline{\delta})}$, if and only if there is no arbitrage. If the GOP exists, its value process is unique.*

The necessity of no arbitrage is straightforward, as indicated above. The sufficiency will follow directly once the numéraire property of the GOP has been established, see Theorem 1.4 below. In a more general continuous time case, the equivalence between no arbitrage and the existence of a GOP, as predicted from Theorem 1.4, is not completely true and technically much more involved. The uniqueness of the GOP only concerns the value process, not the strategy. If there are redundant assets, the GOP strategy is not necessarily unique. Uniqueness of the value process will follow from the Jensen inequality, once the numéraire property has been established. The existence and uniqueness of a GOP plays only a minor role in the theory of investments, where it is more or less taken for granted. In the line of literature that deals with the application of the GOP for pricing purposes, establishing existence is essential.

It is possible to infer some simple properties of the GOP strategy, without further specifications of the model:

Theorem 1.2. *The GOP strategy has the following properties:*

> *(1) The fractions of wealth invested in each asset are independent of the level of total wealth.*
> *(2) The invested fraction of wealth in asset i is proportional to the return on asset i.*
> *(3) The strategy is myopic.*

The first part is to be understood in the sense that the *fractions* invested are independent of current wealth. Moreover, the GOP strategy allocates funds in proportion to the excess return on an asset. Myopic means short-sighted and implies that the GOP strategy in a given period depends only on the distribution of returns in the next period. Hence, the strategy is independent of the time horizon. Despite the negative connotations the word "myopic" can be given, it may, for practical reasons be quite convenient to have a strategy which only requires the estimation of returns one period ahead. It seems reasonable to assume that return distributions further out in the future are more uncertain. To see why the GOP strategy depends only on the distribution of asset returns one period ahead, note that

$$\mathbb{E}\left[\log(S^{(\delta)}(T))\right] = \log(S^{(\delta)}(0)) + \sum_{i=1}^{T} \mathbb{E}\left[\log(1 + R^{(\delta)}(i))\right].$$

In general, obtaining the strategy in an explicit closed form is not possible. This involves solving a non-linear optimization problem. To see this, the first-order conditions of (1.2) are derived. Since, by Theorem 1.2, the GOP strategy is myopic and the invested fractions are independent of wealth, one needs to solve the problem

$$\sup_{\delta(t)} \mathbb{E}_t\left[\log\left(\frac{S^{(\delta)}(t+1)}{S^{(\delta)}(t)}\right)\right] \tag{1.3}$$

for each $t \in \{0, 1, \ldots, T-1\}$, where $\mathbb{E}_t$ denotes the conditional expectation with respect to $\mathcal{F}_t$. Using the fractions $\pi_\delta^i(t) = \frac{\delta^{(i)}(t)S^{(i)}(t)}{S^{(\delta)}(t)}$, the problem can be written

$$\sup_{\pi_\delta(t)\in\mathbb{R}^d} \mathbb{E}\left[\log\left(1 + \left(1 - \sum_{i=1}^{n}\pi_\delta^i(t)\right)R^0(t) + \sum_{i=1}^{n}\pi_\delta^i(t)R^i(t)\right)\right]. \tag{1.4}$$

The properties of the logarithm ensure that the portfolio will automatically become admissible. By differentiation, the first order conditions become

$$\mathbb{E}_{t-1}\left[\frac{1 + R^i(t)}{1 + R^\delta(t)}\right] = 1 \quad i \in \{0, 1, \ldots, n\}. \tag{1.5}$$

This constitutes a set of $d+1$ non-linear equations to be solved simultaneously such that one is a consequence of the others, due to the constraint that $\sum_{i=0}^{d}\pi_\delta^i = 1$. Although these equations do not generally posses an explicit closed-form solution, there are some special cases which can be handled:

Example 1.1. (Betting on events) Consider a one-period model. At time $t = 1$, the outcome of the discrete random variable X is revealed. If the investor bets on this outcome, he receives a fixed number, α, times his original bet, which is normalized to one dollar. If the expected return from betting is negative, the investor would prefer to avoid betting, if possible. Let $A_i = \{\omega | X(\omega) = x_i\}$ be the sets of mutually exclusive possible outcomes, where $x_i > 0$. Some straightforward manipulations provide

$$1 = \mathbb{E}\left[\frac{1 + R^i}{1 + R^{\delta}}\right] = \mathbb{E}\left[\frac{1_{A_i}}{\pi_{\underline{\delta}}^i}\right] = \frac{\mathbb{P}(A_i)}{\pi_{\underline{\delta}}^i},$$

where 1_A is the indicator function of event A, and, hence, $\pi_{\underline{\delta}}^i = \mathbb{P}(A_i)$. Consequently, the growth-maximizer bets proportionally on the probability of the different outcomes.

In the example above, the GOP strategy is easily obtained since there is a finite number of mutually exclusive outcomes and it was possible to bet on any of these outcomes. It can be seen, by extending the example, that the odds for a given event have no impact on the *fraction* of wealth used to bet on the event. In other words, if all events have the same probability, the pay-off, if the event comes true, does not alter the optimal fractions.

Translated into a financial terminology, Example 1.1 illustrates the case when the market is *complete*. The market is complete whenever Arrow–Debreu securities paying one dollar in one particular state of the world can be replicated, and a bet on each event could be interpreted as buying an Arrow–Debreu security. Markets consisting of Arrow–Debreu securities are sometimes referred to as "horse race markets" because only one security, "the winner", will make a pay-off in a given state. (See also Example 4 in Chapter 2 of this volume.) In a financial setting, the securities are most often not modeled as Arrow–Debreu securities.

Example 1.2. (Complete Markets) Again, a one-period model is considered. Assume that the probability space Ω is finite, and, for $\omega_i \in \Omega$, there is a strategy δ_{ω_i} such that, at time 1,

$$S^{(\delta_{\omega_i})}(\omega) = 1_{(\omega = \omega_i)}.$$

Then, the GOP, by the example above, is to hold a fraction of total wealth equal to $\mathbb{P}(\omega)$ in the portfolio $S^{(\delta_{\omega})}$. In terms of the original securities, the investor needs to invest

$$\pi^i = \sum_{\omega \in \Omega} \mathbb{P}(\omega)\pi_{\delta_{\omega}}^i,$$

where $\pi_{\delta_{\omega}}^i$ is the fraction of asset i held in the portfolio $S^{(\delta_{\omega})}$.

The conclusion that a GOP can be obtained explicitly in a complete market is quite general. In an incomplete discrete time setting, the situation is more complicated and no explicit solution will exist, requiring the use of numerical methods to solve the non-linear first-order conditions. The non-existence of an explicit solution to the problem was mentioned by, e.g., [Mossin (1972)] as a main reason for the lack of popularity of the Growth Optimum model in the seventies. Due to the increase in computational power over the past thirty years, time considerations have become unimportant. Leaving the calculations aside for a moment, let us turn to the distinguishing properties of the GOP, which have made it quite popular among academics and investors searching for a utility-independent criterion for portfolio selection. A discussion of the role of the GOP in asset allocation and investment decisions is postponed to Section 1.3.

Theorem 1.3. *The portfolio process $S^{(\delta)}(t)$ has the following properties:*

> *(1) If assets are infinitely divisible, the ruin probability, $\mathbb{P}(S^{(\delta)}(t) = 0$ for some $t \le T)$, of the GOP is zero.*
>
> *(2) If, additionally, there is at least one asset with nonnegative expected growth rate, then the long-term ruin probability (defined below) of the GOP is zero.*
>
> *(3) For any strategy δ it holds that $\limsup \frac{1}{t} \log \left(\frac{S^{(\delta)}(t)}{S^{(\delta)}(t)} \right) \le 0$, almost surely.*
>
> *(4) Asymptotically, the GOP maximizes median wealth.*

The no-ruin property critically depends on *infinite divisibility* of investments. This means that an arbitrary small amount of a given asset can be bought or sold. As wealth becomes low, the GOP will require a constant fraction to be invested and, hence, such a low absolute amount must be feasible. If not, ruin is a possibility. In general, any strategy which invests a fixed relative amount of capital will never cause the ruin of the investor in finite time as long as arbitrarily small amounts of capital can be invested. In the case where the investor is guaranteed not to be ruined at some fixed time, the *long-term ruin probability* of an investor following the strategy δ is defined as

$$\mathbb{P}(\liminf_{t \to \infty} S^{(\delta)}(t) = 0).$$

Only if the optimal growth rate is greater than zero can ruin, in this sense, be avoided. Note that seemingly rational strategies such as "bet such that $\mathbb{E}[X_t]$ is maximized" can be shown to ensure certain ruin, even in fair or

favorable games. A simple example would be head or tail using a false
coin, where chances of a head are 90%. If a player bets all his money on
heads, then the chance that he will be ruined in n games will be $1 - 0.9^n \rightarrow$
1. Interestingly, certain portfolios selected by maximizing utility can have
a long-term ruin probability of one, even if there exist portfolios with a
strictly positive growth rate. This means that some utility-maximizing
investors are likely to end up with, on average, very little wealth. The
third property is the distinguishing feature of the GOP. It implies that, with
probability one, the GOP will overtake the value of any other portfolio and
stay ahead indefinitely. In other words, for *every* path taken, if the strategy
δ is different from the GOP, there is an instant s such that $S^{(\hat{\delta})}(t) > S^{(\delta)}(t)$
for every $t > s$. Hence, although the GOP is defined so as to maximize the
expected growth rate, it also maximizes the long-term growth rate in an
almost sure sense. The proof in a simple case is due to [Kelly (1956)]; more
sources are cited in the notes. This property has led to some confusion; if
the GOP outperforms any other portfolio at some point in time, it may be
tempting to argue that long-term investors should all invest in the GOP.
This is, however, not literally true and this will be discussed in Section
1.3.1. The last part of the theorem has received less attention. Since the
median of a distribution is unimportant to an investor maximizing expected
utility, the fact that the GOP maximizes the median of wealth in the long
run is of little theoretical importance, at least in the field of economics.
Yet, for practical purposes, it may be interesting, since, for highly-skewed
distributions the median is quite useful as a measure of the most likely
outcome. The property was recently shown by [Ethier (2004)].

Another performance criterion often discussed is the expected time to
reach a certain level of wealth. In other words, if the investor wants to
get rich fast, what strategy should he use? It is *not* generally true that
the GOP is the strategy which minimizes this time, due to the problem of
overshooting. If one uses the GOP, the chances are that the target level
is exceeded significantly. Hence a more conservative strategy might be
better, if one wishes to attain a goal and there is no "bonus" for exceeding
the target. To give a mathematical formulation, define

$$\tau^\delta(x) \triangleq \inf\{t \mid S^{(\delta)}(t) \geq x\}$$

and let $g^\delta(t)$ denote the growth rate of the strategy δ, at time $t \in \{1, \dots, \}$.
Note that, due to myopia, the GOP strategy does not depend on the final
time, so it makes sense to define it even if $T = \infty$. Hence, $g^{\underline{\delta}}(t)$ denotes
the expected growth rate using the GOP strategy. If returns are i.i.d., then

$g^{\underline{\delta}}(t)$ is a constant, $g^{\underline{\delta}}$. Defining the stopping time $\tau^{(\delta)}(x)$ to be the first time the portfolio $S^{(\delta)}$ exceeds the level x, the following asymptotic result holds true.

Lemma 1.1. [Breiman (1961)] *Assume returns to be i.i.d. Then, for any strategy δ,*

$$\lim_{x \to \infty} \left(\mathbb{E}[\tau^{(\delta)}(x)] - \mathbb{E}[\tau^{(\underline{\delta})}(x)] \right) = \sum_{i \in \mathbb{N}} \left(1 - \frac{g^{\delta}(i)}{g^{\underline{\delta}}} \right).$$

In fact, a technical assumption needed is that the variables $\log(g^{(\delta)}(t))$ be *non-lattice*. A random variable X is lattice if there is some $a \in \mathbb{R}$ and some $b > 0$ such that $\mathbb{P}(X \in a + b\mathbb{Z}) = 1$, where $\mathbb{Z} = \{\ldots, -2, -1, 0, 1, 2, \ldots\}$. As $g^{\underline{\delta}}$ is larger than g^{δ}, the right-hand side is nonnegative, implying that the expected time to reach a goal is asymptotically minimized when using the GOP, as the desired level is increased indefinitely. In other words, for "high" wealth targets, the GOP will minimize the expected time to reach this target. Note that the assumption of i.i.d. returns implies that the expected growth rate is identical for all periods. For finite hitting levels, the problem of overshooting can be dealt with by introducing a "time rebate" when the target is exceeded. In this case, the GOP strategy remains optimal for finite levels. The problem of overshooting is eliminated in the continuous-time diffusion case, because the diffusion can be controlled instantaneously and, in this case, the GOP will minimize the time to reach any goal; see [Pestien and Sudderth (1985)].

This ends the discussion of the properties that are important when considering the GOP as an investment strategy. Readers whose main interest is in this direction may skip the remainder of this section. Apart from the growth property there is another property, of the GOP, the *numéraire property*, which will be explained below, and which is important in order to understand the role of the GOP in the fields of derivative/asset pricing. Consider Equation (1.5) and assume there is a solution satisfying these first-order conditions. It follows immediately that the resulting GOP will have the property that expected returns of any asset measured against the return of the GOP will be zero. In other words, if GOP-denominated returns of any portfolio are zero, then GOP-denominated prices become *martingales*, since

$$\mathbb{E}_t \left[\frac{1 + R^{\delta}(t+1)}{1 + R^{\underline{\delta}}(t+1)} \right] = \mathbb{E}_t \left[\frac{S^{(\delta)}(t+1)}{S^{(\delta)}(t+1)} \frac{S^{(\underline{\delta})}(t)}{S^{(\underline{\delta})}(t)} \right] = 1,$$

which implies that

$$\mathbb{E}_t \left[\frac{S^{(\delta)}(t+1)}{S^{(\hat{\delta})}(t+1)} \right] = \frac{S^{(\delta)}(t)}{S^{(\hat{\delta})}(t)}.$$

If asset prices in GOP-denominated units are martingales, then the empirical probability measure $\mathbb{P}$ is an equivalent martingale measure (EMM). This suggests a way of pricing a given pay-off. Measure it in units of the GOP and take the ordinary average. In fact, this methodology was suggested recently and will be discussed in Section 1.4. Generally, there is no guarantee that (1.5) has a solution. Even if Theorem 1.1 ensures the existence of a GOP, it may be that the resulting strategy does not satisfy (1.5). Mathematically, this is just the statement that an optimum need not be attained in an inner point, but can be attained at the boundary. Even in this case, something may be said about GOP-denominated returns – they become *strictly negative* – and the GOP-denominated price processes become strict supermartingales.

Theorem 1.4. *The process* $\hat{S}^{(\delta)}(t) \triangleq \frac{S^{(\delta)}(t)}{S^{(\hat{\delta})}(t)}$ *is a supermartingale. If* $\pi^{\hat{\delta}}(t)$ *belongs to the interior of the set*

$$\{x \in \mathbb{R}^d | Investing \ the \ fractions \ x \ at \ time \ t \ is \ admissible\},$$

then $\hat{S}^{(\delta)}(t)$ *is a true martingale.*

Note that $\hat{S}^{(\delta)}(t)$ can be a martingale even if the fractions are not in the interior of the set of admissible strategies. This happens in the (rare) cases where the first-order conditions are satisfied on the boundary of this set. The fact that the GOP has the numéraire property follows by applying the bound $\log(x) \le x - 1$ and the last part of the statement is obtained by considering the first-order conditions for optimality, see Equation (1.5). The fact that the numéraire property of the portfolio $S^{(\delta)}$ implies that $S^{(\hat{\delta})}$ is the GOP is shown by considering the portfolio

$$S^{(\epsilon)}(t) \triangleq \epsilon S^{(\delta)}(t) + (1 - \epsilon)S^{(\hat{\delta})}(t),$$

using the numéraire property and letting ϵ tend to zero.

The martingale condition has been used to establish a theory for pricing financial assets, see Section 1.4, and to test whether a given portfolio is the GOP, see Section 1.5. Note that the martingale condition is equivalent to the statement that returns denominated in units of the GOP become zero. A portfolio with this property was called a *numéraire portfolio* by [Long (1990)]. If one restricts the definition such that a numéraire portfolio only covers the case where such returns are exactly zero, then a numéraire portfolio need not exist. In the case where (1.5) has no solution, there is no numéraire portfolio, but under the assumption of no arbitrage there is a GOP and hence the existence of a numéraire portfolio is not a consequence of no arbitrage. This motivated the generalized definition of a numéraire portfolio, made by [Becherer (2001)], who defined a numéraire portfolio as a portfolio, $S^{(\delta)}$, such that for all other strategies, δ, the process $\frac{S^{(\delta)}(t)}{S^{(\underline{\delta})}(t)}$ would be a supermartingale. By Theorem 1.4 this portfolio is the GOP.

It is important to check that the numéraire property is valid, since otherwise the empirical tests of the martingale restriction implied by (1.5) become invalid. Moreover, using the GOP and changing numéraire technique for pricing derivatives becomes unclear as will be discussed in Section 1.4.

A simple example illustrates the situation that GOP-denominated asset prices may be supermartingales.

Example 1.3. (cf. [Becherer (2001)] and [Bühlmann and Platen (2003)]) Consider a simple one-period model and let the market $(S^{(0)}, S^{(1)})$ be such that the first asset is risk-free, $S^{(0)}(t) = 1$, $t \in \{0, T\}$. The second asset has a log-normal distribution $\log(S^{(1)}(T)) \sim \mathcal{N}(\mu, \sigma^2)$ and $S^{(1)}(0) = 1$. Consider an admissible strategy $\delta = (\delta^{(0)}, \delta^{(1)})$ and assume the investor has one unit of wealth. Since

$$S^{(\delta)}(T) = \delta^{(0)} + \delta^{(1)} S^{(1)}(T) \geq 0$$

and $S^{(1)}(T)$ is log-normal, it follows that $\delta^{(i)} \in [0, 1]$ in order for the wealth process to be nonnegative. Now,

$$\mathbb{E}\left[\log\left(S^{(\delta)}(T)\right)\right] = \mathbb{E}\left[\log\left(1 + \delta^{(1)}(S^{(1)}(T) - S^{(0)}(T))\right)\right].$$

First-order conditions imply that

$$\mathbb{E}\left[\frac{S^{(1)}(T)}{1 + \underline{\delta}^{(1)}(S^{(1)}(T) - S^{(0)}(T))}\right] = \mathbb{E}\left[\frac{S^{(0)}(T)}{1 + \underline{\delta}^{(1)}(S^{(1)}(T) - S^{(0)}(T))}\right] = 1.$$

It can be verified that there is a solution to this equation if and only if $|\mu| \leq \frac{\sigma^2}{2}$. If $\mu - \frac{\sigma^2}{2} \leq 0$ then it is optimal to invest everything in $S^{(0)}$. The intuition is that, compared to the risk-less asset, the risky asset has a negative growth rate. Since the two are independent, it is optimal not to invest in the risky asset at all. In this case,

$$\hat{S}^{(0)}(T) = 1, \quad \hat{S}^{(1)}(T) = S^{(1)}(T).$$

It follows that $\hat{S}^{(0)}$ is a martingale, whereas $\hat{S}^{(1)}(T) = S^{(1)}(T)$ is a strict supermartingale, since $\mathbb{E}[S^{(1)}(T)|\mathcal{F}_0] \leq S^{(1)}(0) = 1$. Conversely, if $\mu \geq \frac{\sigma^2}{2}$, then it is optimal to invest everything in asset 1, because the growth rate of the risk-free asset *relative* to the growth rate of the risky asset is negative. The word relative is important because the growth rate in absolute terms is zero. In this case,

$$\hat{S}^{(0)}(T) = \frac{1}{S^{(1)}(T)}, \quad \hat{S}^{(1)}(T) = 1$$

and, hence, $\hat{S}^{(0)}$ is a supermartingale, whereas $\hat{S}^{(1)}$ is a martingale.

The simple example shows that there is economic intuition behind the case when GOP-denominated asset prices become true martingales. It happens in two cases. Firstly, it may happen if the growth rate of the risky asset is low. In other words, the market price of risk is very low and investors cannot create short positions due to limited liability to short the risky asset. Secondly, it may happen if the risky asset has a high growth rate, corresponding to the situation where the market price of risk is high. In the example, this corresponds to $\mu \geq \frac{\sigma^2}{2}$. Investors cannot have arbitrary long positions in the risky assets, because of the risk of bankruptcy. The fact that investors avoid bankruptcy is not a consequence of Definition 1.1; it will persist even without this restriction. Instead, it derives from the fact that the logarithmic utility function turns to minus infinity as wealth turns to zero. Consequently, any strategy that may result in zero wealth with positive probability will be avoided. One may expect to see the phenomenon in more general continuous-time models, in cases where investors are facing portfolio constraints or if there are jumps which may suddenly reduce the value of the portfolio. We will return to this issue in the next section.

Notes

The assumption of independent returns can be loosened, see [Hakansson and Liu (1970)] and [Algoet and Cover (1988)]. Although strategies in such cases should be based on previous information, not just the information of the current realizations of stock prices, it can be shown that the growth and numéraire property remain intact in this case.

That no arbitrage is necessary seems to have been noted quite early by [Hakansson (1971a)], who formulated this as a "no easy money" condition, where "easy money" is defined as the ability to form a portfolio whose return dominates the risk-free interest rate almost surely. The one-to-one

relation to arbitrage appears in [Maier *et al.* (1977b)][Theorem 1 and 1']
and, although they do not mention arbitrage and state price densities (SPD)
explicitly, their results could be phrased as the equivalence between the
existence of a solution to problem 1.2 and the existence of an SPD [Theorem
1] and the absence of arbitrage [Theorem 1']. The first time the relation is
mentioned explicitly is in [Long (1990)]. Long's Theorem 1, as stated, is *not*
literally true, although it would be if numéraire portfolio was replaced by
GOP. Uniqueness of the value process, $S^{(\delta)}(t)$, was remarked in [Breiman
(1961)][Proposition 1].

The properties of the GOP strategy, in particular the myopia, were analyzed in [Mossin (1968)]. Papers addressing the problem of obtaining a solution to the problem include [Bellman and Kalaba (1957)], [Ziemba (1972)],
[Elton and Gruber (1974)], [Maier *et al.* (1977b)] and [Cover (1984)]. The
methods are either approximations or based on non-linear optimization
models.

The proof of the second property of Theorem 1.3 dates back to [Kelly
(1956)] for a very special case of Bernoulli trials but was noted independently by [Latané (1959)]. The results where refined in [Breiman (1960,
1961)] and extended to general distributions in [Algoet and Cover (1988)].

The expected time to reach a certain goal was considered in [Breiman
(1961)] and the inclusion of a rebate in [Aucamp (1977)] implies that the
GOP will minimize this time for finite levels of wealth.

The numéraire property can be derived from the proof of [Breiman
(1961)][Theorem 3]. The term numéraire portfolio is from [Long (1990)].
The issue of supermartingality was apparently overlooked until explicitly
pointed out in [Kramkov and Schachermayer (1999)][Example 5.1]. A general treatment which takes this into account is found in [Becherer (2001)],
see also [Korn and Schäl (1999, 2009)] and [Bühlmann and Platen (2003)]
for more in a discrete-time setting.

1.2.2. *Continuous-Time*

In this section some of the results are extended to a general continuous-time
framework. The main conclusions of the previous section stand, although
with some important modifications, and the mathematical exposition is
more challenging. For this reason, the results are supported by examples.
Most conclusions from the continuous case are important for the treatment
in Section 1.4 and the remainder of this chapter which is held in continuous-
time.

The mathematical object used to model the financial market given by (1.1) is now a $d + 1$-dimensional semimartingale, S, living on a filtered probability space $(\Omega, \mathcal{F}, \underline{\mathcal{F}}, \mathbb{P})$, satisfying the usual conditions, see [Protter (2004)]. Being a semimartingale, S can be decomposed as

$$S(t) = A(t) + M(t),$$

where A is a finite variation process and M is a local martingale. The reader is encouraged to think of these as *drift* and *volatility*, respectively, but should be aware that the decomposition above is not always unique. If A can be chosen to be predictable, then the decomposition is unique. This is exactly the case when S is a special semimartingale, see [Protter (2004)]. Following standard conventions, the first security is assumed to be the numéraire, and hence it is assumed that $S^{(0)}(t) = 1$, almost surely, for all $t \in [0, T]$. The investor needs to choose a strategy, represented by the $d + 1$ dimensional process

$$\delta = \{\delta(t) = (\delta^{(0)}(t), \ldots, \delta^{(d)}(t)), t \in [0, T]\}.$$

The following definition of admissibility is the natural counterpart to Definition 1.1.

Definition 1.4. An admissible trading strategy, δ, satisfies the three conditions:

(1) δ is an S-integrable, predictable process.
(2) The resulting portfolio value $S^{(\delta)}(t) \triangleq \sum_{i=0}^{d} \delta^{(i)}(t) S^{(i)}(t)$ is non-negative.
(3) The portfolio is self-financing, that is $S^{(\delta)}(t) = \int_0^t \delta(s) dS(s)$.

Here, predictability can be loosely interpreted as left-continuity, but more precisely, it means that the strategy is adapted to the filtration generated by all left-continuous $\underline{\mathcal{F}}$-adapted processes. In economic terms, it means that the investor cannot change his portfolio to guard against jumps that occur randomly. For more on this and a definition of integrability with respect to a semimartingale, see [Protter (2004)]. The second requirement is important in order to rule out simple, but unrealistic, strategies leading to arbitrage, for example doubling strategies. The last requirement states that the investor does not withdraw or add any funds. Recall that $\Theta(S)$ denotes the set of nonnegative portfolios, which can be formed using the elements of S. It is often convenient to consider portfolio fractions of wealth, i.e.

$$\pi_\delta = \{\pi_\delta(t) = (\pi_\delta^0(t), \ldots, \pi_\delta^d(t))^\top, \quad t \in [0, \infty)\}$$

with coordinates defined by:

$$\pi_\delta^i(t) \triangleq \frac{\delta^{(i)}(t)S^{(i)}(t)}{S^{(\delta)}(t)}. \tag{1.6}$$

One may define the GOP, $S^{(\delta)}$, as in Definition 1.2, namely as the solution to the problem

$$S^{(\delta)} \triangleq \arg \sup_{S^{(\delta)} \in \underline{\Theta}(S)} \mathbb{E}[\log(S^{(\delta)}(T))]. \tag{1.7}$$

This, of course, only makes sense if the expectation is uniformly bounded on $\underline{\Theta}(S)$, although alternative and economically meaningful definitions exist which circumvent the problem of having

$$\sup_{S^{(\delta)} \in \underline{\Theta}(S)} \mathbb{E}[\log(S^{(\delta)}(T))] = \infty.$$

For simplicity, let us use the following definition.

Definition 1.5. A portfolio is called a GOP if it satisfies (1.7).

In discrete-time, there was a one-to-one correspondence between no arbitrage and the existence of a GOP. Unfortunately, this breaks down in continuous-time. Here several definitions of arbitrage are possible. A key existence result is based on the article [Kramkov and Schachermayer (1999)], who used the notion of *no free lunch with vanishing risk* (NFLVR). The essential feature of NFLVR is the fact that it implies the existence of an equivalent martingale measure, see [Delbaen and Schachermayer (1994, 1998)]. More precisely, if asset prices are locally bounded, the measure is an equivalent local martingale measure and if they are unbounded, the measure becomes an equivalent sigma martingale measure. Here, these measures will all be referred to collectively as EMM.

Theorem 1.5. *Assume that*

$$\sup_{S^{(\delta)} \in \underline{\Theta}(S)} \mathbb{E}[\log(S^{(\delta)}(T))] < \infty$$

and that NFLVR holds. Then, there is a GOP.

Unfortunately, there is no clear one-to-one correspondence between the existence of a GOP and no arbitrage in the sense of NFLVR. In fact, the GOP may easily exist, even when NFLVR is not satisfied, and NFLVR does not guarantee that the expected growth rates are bounded. Moreover, the choice of numéraire influences whether or not NFLVR holds. A less stringent and numéraire invariant condition is the requirement that the market should have a *martingale density*. A martingale density is a strictly positive process Z, such that $\int S dZ$ is a local martingale. In other words, a Radon–Nikodym derivative of some EMM is a martingale density, but a martingale density is only the Radon–Nikodym derivative of an EMM if it is a true martingale. Modifying the definition of the GOP slightly, one may show that:

Corollary 1.1. *There is a GOP if and only if there is a martingale density.*

The reason why this addition to the previous existence result may be important is discussed in Section 1.4.

Finding the growth optimal strategy in the current setting can be a non-trivial task. Before presenting the general result, an important, yet simple, example is presented.

Example 1.4. Let the market consist of two assets, a stock and a bond. Specifically, the SDEs describing these assets are given by

$$dS^{(0)}(t) = S^{(0)}(t) r dt \quad \text{and}$$
$$dS^{(1)}(t) = S^{(1)}(t) \left(a dt + \sigma dW(t) \right),$$

where W is a Wiener process and r, a, σ are constants. Using fractions, any admissible strategy can be written

$$dS^{(\delta)}(t) = S^{(\delta)}(t) \left((r + \pi(t)(a - r)) dt + \pi(t) \sigma dW(t) \right).$$

Applying Itô's lemma to $Y(t) = \log(S^{(\delta)}(t))$ provides

$$dY(t) = \left((r + \pi(t)(a - r) - \frac{1}{2}\pi(t)^2 \sigma^2) dt + \pi(t) \sigma dW(t) \right).$$

Hence, assuming the local martingale with differential $\pi(t)\sigma dW(t)$ to be a true martingale, it follows that

$$\mathbb{E}[\log(S^{(\delta)}(T))] = \mathbb{E}\left[\int_0^T (r + \pi(t)(a - r) - \frac{1}{2}\pi(t)^2 \sigma^2) dt \right],$$

so, by maximizing the expression for each (t, ω), the optimal fraction is obtained as

$$\pi_{\underline{\delta}}(t) = \frac{a - r}{\sigma^2}.$$

Hence, inserting the optimal fractions into the wealth process, the GOP is described by the SDE

$$dS^{(\delta)}(t) = S^{(\delta)}(t) \left(\left(r + \left(\frac{a-r}{\sigma} \right)^2 \right) dt + \frac{a-r}{\sigma} dW(t) \right)$$
$$\triangleq S^{(\delta)}(t) \left((r + \theta^2) dt + \theta dW(t) \right).$$

The parameter $\theta = \frac{a-r}{\sigma}$ is the market price of risk process.

The example illustrates how the myopic properties of the GOP makes it relatively easy to derive the portfolio fractions. Although the method seems heuristic, it will work in very general cases and, when asset prices are continuous, an explicit solution is always possible. This, however, is not true in the general case. A very general result was provided in [Goll and Kallsen (2000, 2003)], who showed how to obtain the GOP in a setting with intermediate consumption and consumption taking place according to a (possibly random) consumption clock. Here, the focus will be on the GOP strategy and its corresponding wealth process, whereas the implications for optimal consumption will not be discussed. In order to state the result, the reader is reminded of the semimartingale *characteristic triplet*, see [Jacod and Shiryaev (1987)]. Fix a truncation function, h, i.e., a bounded function with compact support, $h : \mathbb{R}^d \to \mathbb{R}^d$, such that $h(x) = x$ in a neighborhood around zero. For example, a common choice would be $h(x) = x1_{(|x| \leq 1)}$. For such truncation function, there is a triplet (A, B, ν), describing the behavior of the semimartingale. One may choose a "good version"; that is, there exists a locally integrable, increasing, predictable process, $\hat{A}$, such that (A, B, ν) can be written as

$$A = \int a d\hat{A}, \quad B = \int b d\hat{A}, \quad \text{and} \quad \nu(dt, dv) = d\hat{A}_t F(t, dv).$$

The process A is related to the finite variation part of the semimartingale, and it can be thought of as a generalized drift. The process B is similarly interpreted as the quadratic variation of the continuous part of S, or, in other words, it is the square volatility, where volatility is measured in absolute terms. The process ν is the compensated jump measure, interpreted as the expected number of jumps with a given size over a small interval and F essentially characterizes the jump size. Note that A depends on the choice of truncation function.

Example 1.5. Let $S^{(1)}$ be as in Example 1.4, i.e. geometric Brownian Motion. Then, $\hat{A} = t$ and

$$dA(t) = S^{(1)}(t)adt \quad dB(t) = (S^{(1)}(t)\sigma)^2 dt.$$

Theorem 1.6. [Goll and Kallsen (2000)] *Let S have a characteristic triplet (A, B, ν) as described above. Suppose there is an admissible strategy $\underline{\delta}$ with corresponding fractions $\pi_{\underline{\delta}}$, such that*

$$a^j(t) - \sum_{i=1}^{d} \frac{\pi_{\underline{\delta}}^i(t)}{S^{(i)}(t)} b^{i,j}(t) + \int_{\mathbb{R}^d} \left(\frac{x^j}{1 + \sum_{i=1}^{d} \frac{\pi_{\underline{\delta}}^i(t)}{S^{(i)}(t)} x^i} - h(x) \right) F(t, dx) = 0$$

(1.8)

for $\mathbb{P} \otimes d\hat{A}$ almost all $(\omega, t) \in \Omega \times [0, T]$, where $j \in \{0, \ldots, d\}$ and $\otimes$ denotes the standard product measure. Then $\underline{\delta}$ is the GOP strategy.

Essentially, Equation (1.8) represents the first-order conditions for optimality and they would be obtained easily if one tried to solve the problem in a pathwise sense, as done in Example 1.4. From Example 1.3 in the previous section, it is clear that such a solution need not exist, because there may be a "corner solution".

The following examples show how to apply Theorem 1.6.

Example 1.6. Assume that discounted asset prices are driven by an m-dimensional Wiener process. The locally-risk-free asset is used as numéraire, whereas the remaining risky assets evolve according to

$$dS^{(i)}(t) = S^{(i)}(t)a^i(t)dt + \sum_{j=1}^{m} S^{(i)}(t)b^{i,j}(t)dW^j(t)$$

for $i \in \{1, \ldots, d\}$. Here $a^i(t)$ is the excess return above the risk-free rate. From this equation, the decomposition of the semimartingale S follows directly. Choosing $\hat{A} = t$, a good version of the characteristic triplet becomes

$$(A, B, \nu) = \left(\int a(t)S(t)dt, \int S(t)b(t)(S(t)b(t))^\top dt, 0 \right).$$

Consequently, in vector form and after division by $S^{(i)}(t)$, Equation (1.8) yields that

$$a(t) - (b(t)b(t)^\top)\pi_{\underline{\delta}}(t) = 0.$$

In the particular case where $m = d$ and the matrix b is invertible, we obtain the well-known result that

$$\pi(t) = b^{-1}(t)\theta(t),$$

where $\theta(t) = b^{-1}(t)a(t)$ is the market price of risk.

Generally, whenever the asset prices can be represented by a continuous semimartingale, a closed form solution to the GOP strategy may be found. The cases where jumps are included are less trivial as shown in the following example.

Example 1.7. (Poissonian Jumps) Assume that discounted asset prices are driven by an m-dimensional Wiener process, W, and an $n - m$ dimensional Poisson jump process, N, with intensity $\lambda \in \mathbb{R}^{n-m}$. Define the compensated Poisson process $q(t) \triangleq N(t) - \int_0^t \lambda(s)ds$. Then, asset prices evolve as

$$dS^{(i)}(t) = S^{(i)}(t)a^i(t)dt + \sum_{j=1}^{m} S^{(i)}(t)b^{i,j}(t)dW^j(t) + \sum_{j=m+1}^{n} S^{(i)}(t)b^{i,j}(t)dq^j(t)$$

for $i \in \{1, \ldots, d\}$. If it is assumed that $n = d$, then an explicit solution to the first-order conditions may be found. Assume that $b(t) = \{b^{i,j}(t)\}_{i,j \in \{1,\ldots,d\}}$ is invertible. This follows if it is assumed that no arbitrage exists. Define

$$\theta(t) \triangleq b^{-1}(t)(a^1(t), \ldots, a^d(t))^\top.$$

If $\theta^j(t) \geq \lambda^j(t)$ for $j \in \{m+1, \ldots, d\}$, then there is an arbitrage, so it can be assumed that $\theta^j(t) < \lambda^j(t)$. In this case, the GOP fractions satisfy the equation

$$(\pi^1(t), \ldots, \pi^d(t))^\top$$
$$= (b^\top)^{-1}(t)\left(\theta^1(t), \ldots, \theta^m(t), \frac{\theta^{m+1}(t)}{\lambda^{m+1}(t) - \theta^{m+1}(t)}, \ldots, \frac{\theta^d(t)}{\lambda^d(t) - \theta^d(t)}\right)^\top.$$

It can be seen that the optimal fractions are no longer linear in the market price of risk. This is because when jumps are present, investments cannot be scaled arbitrarily, since a sudden jump may imply that the portfolio becomes nonnegative. Note that the market price of jump risk needs to be less than the intensity for the expression to be well-defined. If the market is complete, then this restriction follows by the assumption of no arbitrage.

In general, when jumps are present, there is no explicit solution in an incomplete market. In such cases, it is necessary to use numerical methods to solve Equation (1.8). As in the discrete case, the assumption of complete markets will enable the derivation of a fully explicit solution of the problem. In the case of more general jump distributions, where the jump measure does not have a countable support set, the market cannot be completed by any finite number of assets. The jump uncertainty which appears in this case can then be interpreted as driven by a Poisson process of an infinite dimension. In this case, one may still find an explicit solution if the definition of a solution is generalized slightly as in [Christensen and Larsen (2007)].

As in discrete-time, the GOP can be characterized in terms of its growth properties.

Theorem 1.7. *The GOP has the following properties:*

(1) The GOP maximizes the instantaneous growth rate of investments.
(2) In the long-term, the GOP will have a higher realized growth rate than any other strategy, i.e.,

$$\limsup_{T \to \infty} \frac{1}{T} \log(S^{(\delta)}(T)) \leq \limsup_{T \to \infty} \frac{1}{T} \log(S^{(\hat{\delta})}(T))$$

for any other admissible strategy $S^{(\delta)}$.

The instantaneous growth rate is the drift of $\log(S^{(\delta)}(t))$.

Example 1.8. In the context of Example 1.4, the instantaneous growth rate, $g^{\delta}(t)$, of a portfolio $S^{(\delta)}$ was found by applying the Itô formula to obtain

$$dY(t) = \left(\left(r + \pi(t)(a - r) - \frac{1}{2}\pi(t)^2\sigma^2 \right) dt + \pi(t)\sigma dW(t) \right).$$

Hence, the instantaneous growth rate is

$$g^{\delta}(t) = r + \pi(t)(a - r) - \frac{1}{2}\pi(t)^2\sigma^2.$$

In Example 1.4 the GOP was derived, exactly by maximizing this expression and so the GOP maximized the instantaneous growth rate by construction.

As mentioned, the procedure of maximizing the instantaneous growth rate may be applied in a straightforward fashion in more general settings. In the case of a Wiener-driven diffusion with deterministic parameters, the second claim can be obtained directly by using the law of large numbers for Brownian motion. The second claim does not rest on the assumption of continuous asset prices, although this was the setting in which it was proved. What is important is that other portfolios measured in units of the GOP become supermartingales. Since this is shown below for the general case of semimartingales, the proof in [Karatzas (1989)] will also apply here, as shown in [Platen (2004a)].

As in the discrete setting, the GOP enjoys the numéraire property. However, there are some subtle differences.

Theorem 1.8. *Let $S^{(\delta)}$ denote any admissible portfolio process and define $\hat{S}^{(\delta)}(t) \triangleq \frac{S^{(\delta)}(t)}{S^{(\hat{\delta})}(t)}$. Then*

(1) $\hat{S}^{(\delta)}(t)$ is a supermartingale if and only if $S^{(\hat{\delta})}(t)$ is the GOP.
(2) The process $\frac{1}{\hat{S}^{(\delta)}(t)}$ is a submartingale.
(3) If asset prices are continuous, then $\hat{S}^{(\delta)}(t)$ is a local martingale.

In the discrete case, it was shown that prices denominated in units of the GOP could become strict supermartingales. In the case of (unpredictable) jumps, this may also happen, practically for the same reasons as before. If there is a large expected return on some asset and a very slim chance of reaching values close to zero, the log-investor is implicitly restricted from taking large positions in this asset, because by doing so he would risk ruin at some point. This is related to the structure of the GOP in a complete market as explained in Example 1.7.

There is a small but important difference to the discrete-time setting. It may be that GOP-denominated prices become *strict local* martingales, which is a local martingale but does not satisfy the martingale property (this terminology was introduced by [Elworthy *et al.* (1999)]). This is a special case of being a strict supermartingale since, due to the Fatou lemma, a nonnegative local martingale may become a supermartingale. This case does not arise because of any implicit restraints on the choice of portfolios and the threat of being illiquid. Instead, it has to do with the fact that not all portfolios "gets the biggest bang for the buck" as will be explained in Section 1.4.

Example 1.9. (Example 1.4 continued) Assume a market as in Example 1.4 and define the processes $\hat{S}^{(0)}$ and $\hat{S}^{(1)}$ as in Theorem 1.8 above. An application of the Itô formula implies that

$$d\hat{S}^{(0)}(t) = -\hat{S}^{(0)}(t)\theta dW(t)$$

and

$$d\hat{S}^{(1)}(t) = \hat{S}^{(1)}(t)(b(t) - \theta)dW(t).$$

The processes above are local martingales since they are Itô integrals with respect to a Wiener process. A sufficient condition for a local martingale to be a true martingale is given by the so-called Novikov condition, see [Novikov (1973)] requiring

$$\mathbb{E}\left[\exp\left(\frac{1}{2}\int_0^T \theta(t)^2 dt\right)\right] < \infty,$$

which is satisfied in this case since θ is a constant. However, in more general models, θ can be a stochastic process. Several examples exist where the Novikov condition is not satisfied and hence the processes $\hat{S}^{(0)}$ and $\hat{S}^{(1)}$ become true supermartingales. A simple example is the situation where $S^{(1)}$ is a Bessel process of dimension three. The inverse of this process is the standard example of a local martingale which is not a martingale.

The fact that local martingales need not be martingales is important in the theory of arbitrage free pricing and will be discussed in Section 1.4. In these cases, the numéraire may be outperformed by trading the GOP.

The growth properties indicating that, in the long run the GOP will outperform all other portfolios, have made it very interesting in the literature on asset allocation and it has been argued that the GOP is a universally "best" investment strategy in the long run. This application and the debate it has raised in the academic community are reviewed in the next section. A second and more recent application is the numéraire property, particularly interesting in the literature on arbitrage pricing, are reviewed subsequently.

Notes

The literature on the properties of the GOP is huge and only a few references have been discussed here. The properties of this section have been selected because they have attracted the most interest in the literature. Using the logarithm as a utility function often provides very tractable results, so the GOP arises implicitly in a large number of papers which, for simplicity, use this function as part of the theory. To manage the literature on the subject, we have focused on papers which deal explicitly with the GOP. Theorem 1.5 appears in [Becherer (2001)][Theorem 4.5] and is a straightforward application of [Kramkov and Schachermayer (1999)][Theorem 2.2]. In some papers the GOP is defined in a pathwise sense, see [Platen (2002, 2004b)] and [Christensen and Platen (2005)], which circumvents the problem of infinite expected growth rates. An alternative solution is to define the GOP in terms of relative growth rates, see [Algoet and Cover (1988)]. An alternative existence proof, which is more direct, but does not relate explicitly to the notion of arbitrage, can be found in [Aase (1988)] and [Aase and Oksendal (1988)]. [Long (1990)][Appendix B, page 58] claims that the existence of the GOP follows from no arbitrage alone, but this is, in general, incorrect. The proof that the existence of a GOP is equivalent to the existence of a martingale density is found in [Christensen and Larsen (2007)].

Theorem 1.6 was proved by [Goll and Kallsen (2000)] and expanded to stochastic consumption clocks in [Goll and Kallsen (2003)]. The solution in a complete Wiener-driven case with constant parameters dates back to [Merton (1969)], extended in [Merton (1971, 1973)]. [Aase (1988)] introduced the problem in a jump-diffusion setting and derived a similar formula in the context of a model with Wiener and Poisson noise using the Bellman principle. This has been extended in [Aase (1984, 1986, 1988)], [Browne (1999)], [Korn et al. (2003)]. [Yan et al. (2000)] and [Hurd (2004)] study exponential Levy processes and [Platen (2004b)] obtains a fully-explicit solution in the case of a complete Poisson/Wiener market, similar to Example 1.7. It was noted by [Aase (1984)] that Equation (1.8) would follow from a pathwise optimization problem. [Christensen and Platen (2005)] follows this procedure in a general marked-point process setting and expresses the solution in terms of the market price of risk. It is shown that a generalized version of the GOP can be characterized explicitly and approximated by a sequence of portfolios in approximately complete markets. In an abstract framework, relying on the duality result of [Kramkov and Schachermayer (1999)] and the decomposition of [Schweizer (1995)], a general solution was obtained in [Christensen and Larsen (2007)].

The problem of determining the GOP can be extended to the case of portfolio constraints; see, for example, [Cvitanić and Karatzas (1992)] and in particular [Goll and Kallsen (2003)]. The case of transaction costs is considered in [Serva (1999)], [Cover and Iyengar (2000)] and [Aurell and Muratore-Ginanneschi (2004)]. Cases where the growth optimizer has access to a larger filtration are treated by, for example, [Ammendinger et al. (1998)], who show how expanding the set of available information increases the maximal growth rate. In the setting of continuous asset prices, [Larsen and Zitković (2008)] show that the existence of a GOP when the filtration is enlarged guarantees that the price process will remain a semimartingale, which is convenient since arbitrage may arise in models where this property is not guaranteed. A model-free approach to the maximization of portfolio growth rate is derived in [Cover (1991)], and the literature on "universal portfolios".

Theorem 1.7 Property (1) has often been used as the definition of the GOP. Property (2) was proved in [Karatzas (1989)] in the setting of continuous diffusions. For further results on the theoretical long-term behavior of the GOP in continuous-time, the reader is referred to [Pestien and Sudderth (1985)], [Heath *et al.* (1987)] and [Browne (1999)]. Analysis of the long-term behavior and ruin probability is conducted in [Aase (1986)].

The numéraire property in continuous-time was shown initially by [Long (1990)]. The issue of whether GOP-denominated prices become supermartingales is discussed in [Becherer (2001)], [Korn *et al.* (2003)], [Hurd (2004)] and [Christensen and Larsen (2007)]. The fact that the GOP is a submartingale in any other denomination is shown, by for example, [Aase (1988)]. For examples of models, where the inverse GOP is not a true local martingale, see [Delbaen and Schachermayer (1995a)] for a very simple example and [Heath and Platen (2002a)] for a more elaborate one. The standard mathematical textbook reference for such processes is [Revuz and Yor (1991)].

1.3. The GOP as an Investment Strategy

When the GOP was introduced to the finance community, it was not as the result of applying a logarithmic utility function, but in the context of maximizing growth. Investment theory based on growth is an alternative to utility theory and is directly applicable because the specification of the goal is quite simple. The popularity of the mean-variance approach is probably not to be found in its theoretical foundation, but rather the fact that it suggested a simple trade-off between return and uncertainty. Mean-variance based portfolio choice left one free parameter, the amount of variance acceptable to the individual investor. The theory of growth-optimal investment suggests the GOP as an investment tool for long-horizon investors because of the properties stated in the previous section, in particular because it will almost surely dominate other investment strategies in terms of wealth as the time horizon increases. Hence, in the literature of portfolio management, the GOP has often been, and is still, advocated as a useful investment strategy, because utility maximization is a somewhat abstract investment goal. For example, [Roy (1952)][Page 433] states that

> "In calling in a utility function to our aid, an appearance of generality is achieved at the cost of a loss of practical significance, and applicability in our results. A man who seeks advice about his actions will not be grateful for the suggestion that he maximize expected utility."

In these words lies the potential strength of the growth-optimal approach to investment. However, utility theory, being a very influential if not *the* dominating paradigm, is a challenge to alternative, normative, theories of portfolio selection. If investors are correctly modeled as individuals who maximize some (non-logarithmic) utility function, then the growth rate *per se* is of no importance and it makes no sense to recommend the GOP to such individuals.

In this section three issues will be discussed. Firstly, the lively debate on how widely the GOP can be applied as an investment strategy is reviewed in detail. This debate contains several points which may be useful to keep in mind, since new papers in this line of literature often express the point of view that the GOP deserves a special place in the universe of investment strategies. In Section 1.3.1, the discussion of whether the GOP can replace or proxy other investment strategies when the time horizon of the investor is long is presented. The section is aimed to be a chronological review of the pros and cons of the GOP as seen by different authors. Secondly, because the strategy of maximizing growth appeared as a challenge to the well-established mean-variance dogma, and because a large part of the literature has compared the two, Section 1.3.2 will deal with the relation between growth-optimal investments and mean-variance-efficient investments. Finally, because the main argument for the GOP has been its growth properties, some theoretical insight into the ability of the GOP to dominate other strategies over time will be provided in Section 1.3.3.

Before commencing, let me mention that the GOP has found wide applications in gambling and, to some extent, horse racing. In these disciplines, a main issue is how to "gain an edge", i.e. to create a favorable game with nonnegative expected growth rate of wealth. Obviously, if the game cannot be made favorable, i.e. if there is no strategy such that the expected pay-off is larger than the bet, the growth-optimal strategy is of course simply to walk away. If, on the other hand, it is possible to turn the game into a favorable game, then applying the growth-optimal strategy is possible. This can be done in, e.g., Blackjack since simple card-counting strategies can be applied to shift the odds slightly. Similarly, this may be done in horse-racing by playing different bookmakers, see [Hausch and Ziemba (1990)]. There are literally hundreds of papers on this topic. Growth-maximizing strategies are, in this stream of literature, predominantly denoted "Kelly strategies". It appears that Kelly strategies or fractional Kelly strategies are quite common in the theory of gambling and, despite the striking similarity with investment decisions, the gambling literature appears to pay limited attention to the expected utility paradigm in general, perhaps because gamblers, by nature, are much less risk averse than "common investors". A general finding, which may be interesting in the context of asset allocation, is that model uncertainty generally leads to over-betting. Hence, if one wishes to maximize the growth rate of investment, one might wish to apply a fractional Kelly strategy, because the model indicated strategy could be "too risky".

Notes

Some further references for applying the GOP in gambling can be found in [Blazenko *et al.* (1992)], and, in particular, the survey [Hakansson and Ziemba (1995)] and the paper [Thorp (2000)]. See also the papers [Ziemba (2003, 2004)] for some easy-to-read accounts. A standard reference in gambling is the book [Thorp (1966)], while the book [Poundstone (2005)] and the edited volume [Maclean *et al.* (2010)] contain popular treatment of the application of Kelly strategies in gambling and investment.

1.3.1. *Is the GOP Better? – The Samuelson Controversy*

The discussion in this section is concerned with whether the different attributes of the growth-optimal investment constitute reasonable criteria for selecting portfolios. More specifically, we discuss whether the GOP can be said to be "better" in any strict mathematical sense and whether the GOP is an (approximately) optimal decision rule for investors with a long-time horizon. Due to the chronological form of this section and the extensive use of quotes, most references are given in the text, but further references may be found in the notes.

It is a fact that the GOP attracted interest primarily due to the properties stated in Theorem 1.3. A strategy, which in the long run will beat any other strategy in terms of wealth, sounds intuitively attractive, in particular to the investor who is not concerned with short term fluctuations, but has a long-time horizon. Such an investor can lean back and watch his portfolio grow and eventually dominate all others. From this point of view it may sound as if any investor would prefer the GOP, if only his investment horizon is sufficiently long.

Unfortunately, things are not this easy as was initially pointed out by [Samuelson (1963)]. Samuelson argues in his 1963 paper, that, if one is not willing to accept one bet, then one will never rationally accept a sequence of that bet, no matter the probability of winning. In other words, if one does not follow the growth-optimal strategy over one-period, then it will not be rational to follow the rule when there are many periods. His article is not addressed directly to anyone in particular, rather it is written to "dispel a fallacy of wide currency", see [Samuelson (1963)][p. 50]. However, whether it was intended or not, Samuelson's paper serves as a counterargument to the proposed strategy in [Latané (1959)]. Latané had suggested as the criteria for portfolio choice, see [Latané (1959)][p. 146], that one chooses

"...the portfolio that has a greater probability (P') of being as valuable or more valuable than any other significantly different portfolio at the end of n years, n being large."

Latané had argued that this was logical long-term goal, but that it "would not apply to one-in-a-lifetime choices" [p. 145]. This view is repeated in [Latané and Tuttle (1967)]. It would be reasonable to assume that this is the target of Samuelson's critique. Indeed, Samuelson argues that to use this goal is counter-logical, first of all because it does not provide a transitive ordering and, secondly, as indicated above, it is not rational to change objective just because the investment decision is repeated in a number of periods. This criticism is valid to a certain extent, but it is based on the explicit assumption that "acting rationally" means maximizing an expected utility of a certain class. Samuelson's statement is meant as a normative statement. Experimental evidence shows that investors may act inconsistently; see, for example, [Benartzi and Thaler (1999)]. Note that one may construct utility functions, such that two games are accepted, but one is not. An example is in fact given by Samuelson himself in the later paper [Samuelson (1984)]. Further references to this discussion are cited in the notes. However, Latané never claimed his decision rule to be consistent with utility theory. In fact, he seems to be aware of this, as he states

"For certain utility functions and for certain repeated gambles, no amount of repetition justifies the rule that the gamble which is almost sure to bring the greatest wealth is the preferable one."

See [Latané (1959)][footnote 3 on page 145]. [Thorp (1971)] clarifies the argument made by Samuelson that making choices based on the probability that some portfolio will do better or worse than others is non-transitive. However, in the limit, the property characterizing the GOP is that it dominates all other portfolios almost surely. This property, being equally almost surely, clearly is transitive. Moreover, Thorp argues that even in the case where transitivity does not hold, a related form of "approximate

M. M. Christensen

transitivity" does, see [Thorp (1971)][p. 217]. Consequently, he does not argue against Samuelson (at least not directly), but merely points out that the objections made by Samuelson do not pose a problem for his theory. One may wish to emphasize that a comparison of the outcomes as the number of repetitions tends to infinity requires the limit $S^{(\delta)}(t)$ to be well-defined, something which is usually not the case whenever the expected growth rate is nonnegative. However, from Theorem 1.7, the limit

$$\lim_{t \to \infty} \hat{S}^{(\delta)}(t)$$

is well-defined and less than one almost surely. Hence, the question of transitivity depends on whether "n-large" means *in the limit*, in which case it holds or it means for certain *finite* but large n, in which case it does not hold. Second, as pointed out above, "acting rationally" is, in the language of Samuelson, to have preferences that are consistent with a single Von Neumann, Morgenstern utility function. Whether investors who act consistently according to the same utility function ever existed is questionable and this is not assumed by the proponents of the GOP, who intended the GOP as a normative investment rule.

A second question is whether, due to the growth properties, there may be some way to say that "in the long run, everyone should use the GOP".

In this discussion, Samuelson points directly to [Williams (1936)], [Kelly (1956)] and [Latané (1959)]. The main point is that just because the GOP in the long run will end up dominating the value of any other portfolio, it will not be true, over any horizon however long, that the GOP is better for all investors. In Samuelson's own words, see [Samuelson (1971)][p. 2494]:

> "...it is tempting to believe in the truth of the following false corollary:
> *False Corollary.* If maximizing the geometric mean almost certainly leads to a better outcome, then the expected utility of its outcome exceeds that of any other rule, provided that T is sufficiently large."

Such an interpretation of the arguments given by, for example, Latané may be possible, see [Latané (1959)][footnote on page 151]. Later, it becomes absolutely clear that Samuelson did indeed interpret Latané in this way, but otherwise it is difficult to find any statement in the literature which explicitly expresses the point of view which is inherent in the false corollary of [Samuelson (1971)]. Possibly the view point expressed in [Markowitz (1959)] could be interpreted along these lines. Markowitz finds it irrational that long-term investors would not choose the GOP - he does not argue that investors with other utility functions would not do it, but rather he argues that one should not have other utility functions in the very long run. This is criticized by [Thorp (1971)], who points out that the position taken by [Markowitz (1959)] cannot be supported mathematically. Nevertheless, this point of view is somewhat different to that expressed by the false corollary. Whether believers in the false corollary ever existed is questioned by [Thorp (1971)][p. 602]. The point is that one cannot exchange the limits in the following way: if

$$\lim_{t\to\infty} \frac{S^{(\delta)}(t)}{S^{(\underline{\delta})}(t)} \leq 1,$$

then it does not hold that

$$\lim_{t\to\infty} \mathbb{E}[U(S^{(\delta)}(t))] \leq \lim_{t\to\infty} \mathbb{E}[U(S^{(\underline{\delta})}(t))],$$

given some utility function U. This would require, for example, the existence of the pointwise limit $S^{(\underline{\delta})}(\infty)$ and uniform integrability of the random variables $U(S^{(\underline{\delta})}(t))$. Even if the limit and the expectation operator can be exchanged, one might have $\mathbb{E}[U(S^{(\delta)}(t))] > \mathbb{E}[U(S^{(\underline{\delta})}(t))]$ for all finite t and equality in the limit. The intuitive reason is that even if the GOP dominates another portfolio with a very high probability, i.e.

$$\mathbb{P}(S^{(\delta)}(t) < S^{(\underline{\delta})}(t)) = 1 - \epsilon,$$

then the probability of the outcomes where the GOP performs poorly may still be unacceptable to an investor who is more risk averse than a log-utility investor. In other words, the left-tail distribution of the GOP may be too "thick" for an investor who is more risk-averse than the log-utility investor. It seems that a large part of the dispute is caused by claims which argue that the aversion towards such losses is "irrational" because the probability becomes arbitrarily small, whereas the probability of doing better than everyone else becomes large. Whether or not such an attitude is "irrational" is certainly a debatable subject and is probably more a matter of opinion than a matter of mathematics.

When it became clear that the GOP would not dominate other strategies in any crystal-clear sense, several approximation results where suggested. The philosophy was that as the time horizon increased, the GOP would *approximate* the maximum expected utility of other utility functions. However, even this project failed. [Merton and Samuelson (1974a)] pointed out a flaw in an argument in [Hakansson (1971b)] and [Samuelson (1971)], that a log-normal approximation to the distribution of returns over a long period can be made. [Hakansson (1974)] admits to this error, but points out that this has no consequences for the general statements of his paper. Moreover, [Merton and Samuelson (1974a)] remark that a conjecture made by [Samuelson (1971)] and [Markowitz (1972)], that over a long-horizon the GOP will equal or be a good approximation to the optimal policy when investors have bounded utility, is incorrect. Presumably, this unpublished working paper, referred to by [Merton and Samuelson (1974a)] is an older version of [Markowitz (1976)]. Firstly, they remark that [Markowitz (1972)] did not define precisely what a "good approximation" was. Secondly, [Goldman (1974)] gives a counter example showing that following the GOP strategy can lead to a large loss in terms of certainty equivalents, even when investors have a bounded utility function. If U is a bounded utility function, then, certainly, $U(S^{(\delta)}(t))$ is a family of uniformly-integrable variables and, consequently, any converging sequence also converges in mean. This means that it is true that

$$\lim_{T \to \infty} \mathbb{E}[U(S^{(\delta)}(T)] = \lim_{T \to \infty} \mathbb{E}[U(S^{(\underline{\delta})}(T)],$$

but the argument in, for example, [Goldman (1974)], is that $\mathbb{E}[U(S^{(\delta)}(t))]$ converges much slower as $t \to \infty$ than for the optimal policy. In other words, if δ_U is the optimal policy for an investor with utility function U, then [Goldman (1974)] provides an example such that

$$\lim_{t \to \infty} \frac{\mathbb{E}[U(S^{(\underline{\delta})}(t))]}{\mathbb{E}[U(S^{(\delta_u)}(t))]} = 0.$$

So, even though the absolute difference in utility levels when applying the GOP instead of the optimal strategy is shrinking, the GOP performs infinitely worse than the optimal strategy in terms of relative utility. Similarly, one may investigate the certainty equivalent for an investor who is forced to invest in the GOP. The certainty equivalent measures the amount of extra wealth needed to obtain the same level of utility when using a suboptimal strategy. The certainty equivalent when using the GOP in place of the optimal strategy is usually not decreasing as time passes. [Markowitz (1976)] argues that the criterion for asymptotic optimality adopted by [Merton and Samuelson (1974a)] and [Goldman (1974)] is unacceptable, because it violates the notion that only the normalized form of the game is necessary for comparing strategies. The "bribe", which is described as a concept similar to certainty equivalent, cannot be inferred by the normalized form of the game. Markowitz moves on to define utility on a sequence of games and concludes that, if the investor is facing two sequences X and X' and prefers X to X' if $X_n \geq X'_n$ from a certain n with probability one, then such an investor should choose the GOP. A very similar support of the max-expected-growth-rate point of view is given by [Miller (1975)], who shows that, if the utility function depends only on the tail of the wealth sequence of investments, then the GOP is optimal. In technical terms, if $(X_n)_{n\in\mathbb{N}}$ is a sequence such that X_n represents wealth after n periods, then $U : \mathbb{R}^\infty \to \mathbb{R}$ is such that, $U(x_1, \ldots, x_n, \ldots) \geq U(x'_1, \ldots, x'_n, \ldots)$ whenever $x_{n+j} \geq x'_{n+j}$ for some $n \in \mathbb{N}$ and all $j \geq n$. This abstract notion implies that the investor will only care about wealth effects that are "far out in the future". It is unclear whether such a criterion can be given an axiomatic foundation, although it does have some resemblance to the Ramsey–Weizsäcker overtaking criterion used in growth theory, see [Brock (1970)], for the construction of an axiomatic basis.

It seems that the debate on this subject was somewhat obstructed because there was some disagreement about the correct way to measure whether something is "a good approximation". The concept of "the long run" is by nature not an absolute quantity and depends on the context. Hence, the issue of how "long" the "long run" is will be discussed later on.

In the late seventies the discussion became an almost polemic repetition of the earlier debate. [Ophir (1978)] repeats the arguments of Samuleson and provides examples where the GOP strategy, as well as the objective suggested by Latané, will provide unreasonable outcomes. In particular, he notes the lack of transitivity when choosing the investment with the highest probability of providing the best outcome. [Latané (1978)] counter-argues that nothing said so far invalidates the usefulness of the GOP and that he never committed to the fallacies mentioned in Samuelsons paper. As for his choice of objective, Latané refers to the discussion in [Thorp (1971)] regarding the lack of transitivity. Moreover, Latané points out that a goal which he advocates for use when making a long sequence of investment decisions is being challenged by an example involving only one *unique* decision. In [Latané (1959)], Latané puts particular emphasis on the point that goals can be different in the short and long run. As mentioned, this was exactly the reasoning which Samuelson attacks in [Samuelson (1963)]. [Ophir (1979)] refuses to acknowledge that goals should depend on circumstances and once again establishes that Latanés objective is inconsistent with the expected utility paradigm. [Samuelson (1979)] has the last word in his, rather amusing, article which is held in words of only one syllable (apart from the word syllable itself!). In two pages he disputes that the GOP has any special merits, backed by his older papers. The polemic nature of these papers emphasizes that parts of the discussion for and against maximizing growth rates depend on a point of view and is not necessarily supported by mathematical necessities.

To sum up this discussion, there seems to be complete agreement that the GOP can neither dominate nor proxy for other strategies in terms of expected utility, and, no matter how long time a horizon the investor has, utility-based preferences can make other portfolios more attractive because they have a more appropriate risk profile. However, it should be understood that the GOP was recommended as an alternative to expected utility and as a *normative* rather than *descriptive* theory. In other words, authors that argued for the GOP did so because they believed growth-optimality to be a reasonable investment goal, with attractive properties that would be relevant to long-horizon investors. They recommended the GOP because it seems to manifest the desire of accumulating as much wealth as fast as possible. On the other hand, authors who disagreed did so because they did not believe that every investor could be described as a log-utility maximizing investor. Their point is that, if an investor can be *described* as utility maximizing, it is pointless to *recommend* a portfolio which provides less utility than would be the case, should he choose optimally. Hence, the disagreement has its roots in two very fundamental issues, namely whether or not utility theory is a reasonable way of approaching investment decisions in practice and whether the utility function, different from the logarithm, is a realistic description of individual long-term investors. The concept of utility-based portfolio selection, although widely used, may be criticized by the observation that investors may be unaware of their own utility functions. Even the three axioms required in the construction of utility functions, see [Kreps (1988)], have been criticized, because there is some evidence that choices are not made in the coherent fashion suggested by these axioms. Moreover, to say that one strategy provides higher utility than another strategy may be "business as usual" to the economist. Nevertheless, it is a very abstract statement, whose content is based on deep assumptions about the workings of the minds of investors. Consequently, although utility theory is a convenient and consistent theoretical approach, it is not a fundamental law of nature. Neither is it strongly supported by empirical data and experimental evidence. (See, for example, the monograph [Bossaerts (2002)] for some of the problems that asset-pricing theory built on CAPM and other equilibrium models is facing and how some may be explained by experimental evidence on selection.) After the choice of portfolio has been made, it is important to note that only one path is ever realized. It is practically impossible to verify *ex post* whether some given portfolio was "the right choice". By contrast, the philosophy of maximizing growth and the long-run growth property are formulated in dollars, not in terms of utility, and so when one evaluates the portfolio performance *ex post*, there is a greater likelihood that the GOP will come out as a "good idea", because the GOP has a high probability of being more valuable than any other portfolio. It seems plausible that individuals who observe their final wealth will not care that their wealth process is the result of an *ex ante* correct portfolio choice, when it turns out that the performance is only mediocre compared to other portfolios.

Every once in a while, articles continue the debate about the GOP as a very special strategy. These can be separated into two categories. The first, can be represented by [McEnally (1986)] who agrees that the criticism raised by Samuelson is valid. However, he argues that, for practical purposes, in particular when investing for pension, the probability that one will realize a gain is important to investors. Consequently, Latané's subgoal is not without merit in McEnally's point of view. Hence, this category consists of those who simply believe the GOP to be a tool of practical importance and this view reflects the conclusions that have been drawn above.

The second category does not acknowledge the criticism to the same extent and is characterized by statements such as

"... Kelly has shown that repetition of the investment many times gives an objective meaning to the statement that the Growth-optimal strategy is the best, regardless to the subjective attitude to risk or other psychological considerations."

see [Aurell *et al.* (2000b)][Page 4]. The contributions of this specific paper lie within the theory of derivative pricing and will be considered in Section 1.4. Note that they argue in contrary to the conclusions of my previous analysis. In particular, they seem to insist on an interpretation of Kelly, which has been disproved. Their interpretation is even more clear in [Aurell *et al.* (2000a)][Page 5], stating:

"Suppose some agents want to maximize non-logarithmic utility... and we compare them using the growth-optimal strategy, they would almost surely end up with less utility according to their own criterion",

which appears to be a misconception and in general the statement will not hold literally as explained previously. Hence some authors still argue that every *rational* long-term investor should choose the GOP. They seem to believe that either other preferences will yield the same result, which is incorrect, or that other preferences are irrational, which is a viewpoint that is difficult to defend on purely theoretical grounds. A related idea which is sometimes expressed is that it does not make sense to be more risk-seeking than the logarithmic investor. This viewpoint was expressed and criticized very early in the literature. Nevertheless, it seems to have stuck and is found in many papers discussing the GOP as an investment strategy. Whether it is true depends on the context. Although unsupported by utility theory, the viewpoint finds support within the context of growth-based investment. Investors who invest more in risky securities than the fraction warranted by the GOP will, by definition, obtain a lower growth rate over time and, at the same time, they will face more risk. Since the added risk does not imply a higher growth rate of wealth, it constitutes a choice which is "irrational", but only in the same way as choosing a non-efficient portfolio within the mean-variance framework. It is similar to the discussion of whether, in the long run, stocks are better than bonds. In many models, stocks will outperform bonds almost surely as time tends to infinity. Whether long-time-horizon investors should invest more in stocks depends; from utility-based portfolio selection the answer may be no. If the pathwise properties of the wealth distribution are emphasized, then the answer may be yes. As was the case in this section, arguments supporting the last view will often be incompatible with the utility-based theories for portfolio selection. Similar is the argument that "relative risk goes to zero as time goes to infinity" because portfolio values will often converge to infinity as the time horizon increases. Hence, risk measures such as Value at Risk (VaR) will converge to zero as time tends to infinity, which is somewhat counterintuitive, see [Treussard (2005)].

In conclusion, many other unclarities in the finance relate to the fact that a pathwise property may not always be reflected when using expected utility to derive the true portfolio choice. It is a trivial exercise to construct a sequence of random variables that converge to zero, and yet the mean value converges to infinity. In other words, a portfolio may converge to zero almost surely and still be preferred to a risk-free asset by a utility maximizing agent. Intuition dictates that one should never apply such a portfolio over the long-term, whereas the utility maximization paradigm says differently. Similarly, if one portfolio beats others almost surely over a long-time horizon, then intuition suggests that this may be a good investment. Still utility maximization refuses this intuition. It is those highly counterintuitive results which have caused the debate among economists and which continue to cast doubt on the issue of choosing a long-term investment strategy.

As a way of investigating the importance of the growth property of the GOP, Section 1.3.3 sheds light on how long it will take before the GOP moves ahead of other portfolios. We will document that choosing the GOP because it outperforms other portfolios may not be a strong argument, because it may take hundreds of years before the probability of outperformance becomes high.

Notes

The criticism by Samuelson and others can be found in the papers, [Samuelson (1963, 1969, 1971, 1979, 1991)], [Merton and Samuelson (1974a,b)] and [Ophir (1978, 1979)]. The sequence of papers provides a very interesting criticism. Although they do point out certain factual flaws, some of the viewpoints may be characterized as (qualified) opinions rather than truth in any objective sense.

Some particularly interesting references which explicitly take a different stand in this debate is [Latané (1959, 1978)], [Hakansson (1971a,b)] and [Thorp (1971, 2000)], which are all classics. Some recent support is found in [McEnally (1986)], [Aurell *et al.* (2000b)], [Michaud (2003)] and [Platen (2005b)]. The view that investment of more than 100% in the GOP is irrational is common in the gambling literature – referred to as "overbetting" – and is found, for example, in [Ziemba (2003, 2004, 2005)]. In a finance context, the argument is voiced in [Platen (2005c)]. Game-theoretic arguments in favor of using the GOP are found in [Bell and Cover (1980, 1988)]. [Rubinstein (1976)] argues that using generalized logarithmic utility has practical advantages to other utility functions, but does not claim superiority of investment strategies based on such assumptions. The "fallacy of large numbers" problem is considered in numerous papers, for example [Samuelson (1984)], [Ross (1999)], [Brouwer and den Spiegel (2001)] and [Vivian (2003)]. It is shown in [Ross (1999)] that, if utility functions have a bounded first-order derivative near zero, then they may indeed accept a long sequence of bets, while rejecting a single one.

A recent working paper, [Rotar (2004)], considers investors with distorted beliefs, that is, investors who maximize expected utility not with respect to the real world measure, but with respect to some transformation. Conditions such that selected portfolios will approximate the GOP as the time horizon increases to infinity are given.

1.3.2. *Capital Growth and the Mean-Variance Approach*

In the early seventies, the mean-variance approach developed in [Markowitz (1952)] was the dominating theory for portfolio selection. Selecting portfolios by maximizing growth was much less used, but attracted significant attention from academics and several comparisons of the two approaches can be found in the literature from that period. Of particular interest was the question of whether or not the two approaches could be united, or if they were fundamentally different. The conclusion from this investigation along with a comparison of the two approaches will be reviewed here. In general, growth maximization and mean-variance-based portfolio choice are two different things. This is unsurprising, since it is well-known that mean-variance based portfolio selection is not consistent with maximizing a given utility function, except for special cases. Given the theoretically more solid foundation of the growth-optimum theory compared to mean-variance based portfolios selection, we will try to explain why the growth-optimum theory became much less widespread. Most parts of the discussion are presented in discrete-time, but, in the second part of this section, the continuous-time parallel will be considered since the conclusions here are very different.

1.3.2.1. *Discrete time*

Consider the discrete-time framework of Section 1.2.1. Recall that a mean-variance efficient portfolio, is a portfolio such that any other portfolio having the same mean return will have equal or higher variance. These portfolios are obtained as the solution to a quadratic optimization program. It is well-known that the theoretical justification of this approach requires either a quadratic utility function or some fairly restrictive assumption on the class of return distribution, the most common being the assumption of normally distributed returns. The reader is assumed to be familiar with the method, but sources are cited in the notes. Comparing this method for portfolio selection to the GOP yields the following general conclusion.

- The GOP is, in general, not mean-variance efficient. [Hakansson (1971a)] constructs examples such that the GOP lies very far from the efficient frontier. These examples are quite simple and involve only a few assets with two point distributions, but illustrate the fact that the GOP may be far from the mean-variance efficient frontier. This is, perhaps, not surprising, given the fact that mean-variance selection can be related to quadratic utility, whereas growth-optimality is related to logarithmic utility. Only for specific distributions will the GOP be efficient. Note that, if the distribution has support on the entire real axis, then the GOP is trivially efficient, since all money will be put in the risk-free asset. This is the case for normally distributed returns.

- Mean-variance efficient portfolios risk ruin. From Theorem 1.3 and the subsequent discussion, it is known that if the growth rate of some asset is positive and the investment opportunities are infinitely divisible, then the GOP will have no probability of ruin, neither in the short- or the long-run sense. This is not the case for mean-variance efficient portfolios, since there are efficient portfolios which can become negative and some which have a negative expected growth rate. Although portfolios with a negative expected growth rate need not become negative, such portfolios will converge to zero as the number of periods tends to infinity.

- Mean-variance-efficient portfolio choice is inconsistent with first-order stochastic dominance. Since the quadratic-utility function is decreasing from a certain point onwards, a strategy which provides more wealth almost surely may not be preferred to one that brings less wealth. Since the logarithmic function is increasing, the GOP will not be dominated by any portfolio.

The general conclusions above leave the impression that growth-based investment strategies and mean-variance efficient portfolios are very different. This view is challenged by authors who show that approximations of the geometric mean by the first and second moment can be quite accurate. Given the difficulties of calculating the GOP, such approximations were sometimes used to simplify the optimization problem of finding the portfolio with the highest geometric mean, see, for example, [Latané and Tuttle (1967)]. Moreover, the empirical results of Section 1.5 indicate that it can be difficult to tell whether the GOP is in fact mean-variance efficient or not.

In the literature, it has been suggested to construct different trade-offs between growth and security in order for investors with varying degrees of risk aversion to invest more conservatively. These ideas have the same intuitive content as the mean-variance efficient portfolios. One chooses a portfolio which has a desired risk level and which maximizes the growth rate given this restriction. Versions of this trade-off include the compound return mean-variance model, which is, in a sense, a multi-period version of the original one-period mean-variance model. In this model, the GOP is the only efficient portfolio in the long run. More direct trade-offs between growth and security include models where security is measured as the probability of falling short of a certain level, the probability of falling below a certain path, the probability of losing before winning etc.

Interpreted in the context of general equilibrium, the mean-variance approach has been further developed into the well-known CAPM, postulating that the market portfolio is mean-variance efficient. A similar theory was developed for the capital growth criteria by [Budd and Litzenberger (1971)] and [Kraus and Litzenberger (1975)]. If all agents are assumed to maximize the expected logarithm of wealth, then the GOP becomes the market portfolio and, from this, an equilibrium asset pricing model appears. This is not different from what could be done with any other utility function, but the conclusions of the analysis provide empirically-testable predictions and are therefore of some interest. At the heart of the equilibrium model obtained by assuming log-utility is the martingale or numéraire condition. Recall that $R^i(t)$ denotes the return on asset i between time $t-1$ and time t and R^δ is the return process for the GOP. Then, the equilibrium condition is

$$1 = \mathbb{E}_{t-1}\left[\frac{1 + R^i(t)}{1 + R^\delta(t)}\right], \tag{1.9}$$

which is simply the first-order condition for a logarithmic investor. Assume a setting with a finite number of states, that is, $\Omega = \{\omega_1, \ldots, \omega_n\}$, and define $p_i = \mathbb{P}(\{\omega_i\})$. Then, if $S^{(i)}(t)$ is an Arrow–Debreu security, paying off one unit of wealth at time $t+1$, substituting into Equation (1.9) provides

$$S^{(i)}(t) = \mathbb{E}_t\left[\frac{1_{(\omega=\omega_i)}}{1 + R^\delta(t+1)}\right] \tag{1.10}$$

and consequently summing over all states provides an equilibrium condition for the risk-free interest rate

$$1 + r(t+1) = \mathbb{E}_t\left[\frac{1}{1 + R^\delta(t+1)}\right]. \tag{1.11}$$

Combining Equations (1.9) and (1.11), defining $\bar{R}^i \triangleq R^i - r$ and performing some basic, but lengthy manipulations, yield

$$\mathbb{E}_t[\bar{R}^i(t+1)] = \beta_t^i \mathbb{E}_t[\bar{R}^{(\delta)}(t+1)], \tag{1.12}$$

where

$$\beta_t^i = \frac{\text{cov}\big(\bar{R}^i(t+1), \frac{\bar{R}^{\underline{\delta}}(t+1)}{R^{\underline{\delta}}(t+1)}\big)}{\text{cov}\big(\bar{R}^{\underline{\delta}}(t+1), \frac{\bar{R}^{\underline{\delta}}(t+1)}{R^{\underline{\delta}}(t+1)}\big)} .$$

This is similar to the CAPM, apart from the β which in the CAPM has the form

$$\beta_{\text{CAPM}} = \frac{\text{cov}(R^i, R^*)}{\text{var}(R^*)} .$$

In some cases, the CAPM and the CAPM based on the GOP will be very similar. For example, when the characteristic lines are linear or trading intervals the two approaches are indistinguishable and should be perceived as equivalent theories. Later in this section, the continuous-time version will be shown and there the GOP is always instantaneously mean-variance efficient.

Since the growth-based approach to portfolio choice has some theoretically beneficial features compared to the mean-variance theory and the "standard" CAPM, one may wonder why this approach did not find more widespread use. The main reason is presumably the strength of simplicity. Mean-variance-based portfolio choice has an intuitive appeal as it provides a simple trade-off between expected return and variance. This trade-off can be parameterized in a closed form, requiring only the estimation of a variance-covariance matrix of returns and the ability to invert this matrix. Although choosing a portfolio which is either a fractional Kelly strategy or logarithmic mean-variance efficient provides the same trade-off and it is computationally much more involved. In Section 1.2.1 the fact was pointed out that determining the GOP in a discrete-time setting is potentially difficult and no closed form solution is available. Although this may be viewed as a rather trivial matter today, it certainly was a challenge to the computational power available 35 years ago. Second, the theory was attacked immediately for the lack of economic justification. Finally, the empirical data presented in Section 1.5 show that it is very hard to separate the CAPM tangency portfolio and the GOP in practice.

1.3.2.2. *Continuous time*

The assumption that trading takes place continuously is the foundation of the *Intertemporal CAPM* of [Merton (1973)]. Here, the price process of the risky asset $S^{(i)}$, $i \in \{1, \ldots, d\}$, is modeled as a continuous-time diffusion of the form

$$dS^{(i)}(t) = S^{(i)} \left(a^i(t)dt + \sum_{j=1}^{m} b^{i,j}(t)dW^j(t) \right),$$

where W is an m-dimensional standard Wiener process. The process $a^i(t)$ can be interpreted as the instantaneous mean return and $\sum_{j=1}^{m}(b^{i,j}(t))^2$ is the instantaneous variance. One may define the instantaneous mean-variance-efficient portfolios as solutions to the problem

$$\sup_{\delta \in \underline{\Theta}(S)} a^\delta(t)$$
$$\text{s.t. } b^\delta(t) \leq k(t),$$

where $k(t)$ is some nonnegative adapted process. To characterize such portfolios, define the *minimal market price of risk* vector,

$$\theta^p = \{\theta^p(t) = ((\theta^p)^1(t), \ldots, (\theta^p)^m(t))^\top, t \in [0,T]\},$$

by

$$\theta^p(t) \triangleq b(t)(b(t)b(t)^\top)^{-1}(t)(a(t) - r(t)\mathbf{1}). \tag{1.13}$$

Denote the Euclidean norm by

$$\|\theta^p(t)\| = \left(\sum_{j=1}^{m} (\theta^p)^j(t)^2 \right)^{\frac{1}{2}}.$$

Then, instantaneously mean-variance efficient portfolios have fractions which are solutions to the equation

$$(\pi^1(t), \ldots, \pi^N(t))^\top b(t) = \alpha(t)\theta(t) \tag{1.14}$$

for some non-negative process α, and the corresponding SDE for such portfolios is given by

$$dS^{(\delta(\alpha))}(t)$$
$$= S^{(\delta(\alpha))}(t) \left(\left(r(t) + \alpha(t)\|\theta(t)\|^2 \right) dt + \alpha(t) \sum_{j=1}^{m} \theta^j(t)dW^j(t) \right). \tag{1.15}$$

From Example 1.6 it can be verified that, in this case, the GOP is in fact instantaneously mean-variance efficient, corresponding to the choice of $\alpha = 1$. In other words, the GOP belongs to the class of instantaneous

Sharpe ratio maximizing strategies, where the Sharpe ratio, $s^{(\delta)}$, of some strategy δ is defined as

$$s^{(\delta)}(t) = \frac{a^\delta(t) - r(t)}{\sum_{j=1}^{m}(b^{\delta,j}(t))^2}.$$

Here $a^\delta(t) = \delta^{(0)}(t)r(t) + \sum_{i=1}^{n}\delta^{(i)}(t)a^i(t)$ and, similarly, $b^{\delta,j}(t) = \sum_{i=1}^{n}\delta^{(i)}(t)b^{i,j}(t)$.

Note that the instantaneously mean-variance efficient portfolios consist of a position in the GOP and the rest in the risk-free asset, in other words a fractional Kelly strategy. Under certain conditions, for example, if the market price of risk and the interest rate are deterministic processes, it can be shown that any utility-maximizing investor will choose a Sharpe ratio maximizing strategy and, in such cases, fractional Kelly strategies will be optimal for any investor. This result can be generalized to the case where the short rate and the total market price of risk, $||\theta^p(t)||$, are adapted to the filtration generated by the noise source that drives the GOP. It is, however, well-known that if the short rate or the total market price of risk is driven by factors which can be hedged in the market, some investors will choose to do so and consequently not choose a fractional Kelly strategy. When jumps are added to asset prices, the GOP will again become instantaneously mean-variance *inefficient*, except for very special cases. The conclusion is shown to depend strongly on the pricing of event risk and completeness of markets.

If the representative investor has logarithmic utility, then, in equilibrium, the GOP will become the market portfolio. Otherwise, this will not be the case. Since the conditions under which the GOP becomes exactly the market portfolio are thus fairly restrictive, some authors have suggested that the GOP may be very similar to the market portfolio under a set of more general assumptions. For example, it has been shown in [Platen (2003, 2005a)] that sufficiently diversified portfolios will approximate the GOP under certain regularity conditions. (In Chapter 2 of this volume a special mean-variance approach, called semi-log-optimal portfolio, is a good approximation of the GOP.) It should be noted that the circumstances under which the GOP approximates the market portfolio do not rely on the preferences of individual investors. The regularity conditions consist of a limit to the amount of volatility not mirroring that of the GOP. This condition may be difficult to verify empirically.

In the end, whether the GOP is close to the market portfolio and whether the theory based on this assumption holds true remain empirical questions, which will be considered later on. Foreshadowing these conclusions, the general agreement from the empirical analysis is that, if anything, the GOP is more risky than the market portfolio, but rejecting the hypothesis that the GOP is a proxy for the market portfolio is, on the other hand, very difficult.

Notes

The mean-variance portfolio technique is found in most finance textbooks. For proofs and a reasonably rigorous introduction, see [Huang and Litzenberger (1988)]. The main results of the comparison between mean-variance and growth-optimality are found in [Hakansson (1971b,a)]; see also [Hakansson (1974)]. The compound return mean-variance trade-off was introduced in [Hakansson (1971b)]. A critique of this model is found in [Merton and Samuelson (1974a,b)], but some justification is given by [Luenberger (1993)]. Papers discussing the growth-security trade-off include [Blazenko et al. (1992)], [Li (1993)], [Li et al. (2005)], [Michaud (2003)] and [MacLean et al. (2004)]. In the gambling literature, the use of fractional Kelly strategies is widespread. For more references, the reader is referred to [Hakansson and Ziemba (1995)]. An earlier comparison between mean-variance and the GOP is found in [Bickel (1969)]. [Thorp (1969)] recommends to replace the mean-variance criterion with the Kelly-criterion for portfolio selection, due to the sometimes improper choices made by the latter. For approximations of geometric means see [Trent and Young (1969)] and [Elton and Gruber (1974)]. They conclude that the first two moments can provide reasonable approximations, in particular if the distribution does not have very fat tails. In continuous-time, a recent discussion of a growth security trade-off and the CAPM formula which appears can be found in [Bajeux-Besnaino and Portait (1997a)] and [Platen (2002, 2006b, 2005c)]. An application of the GOP for asset-pricing purposes can be found in [Ishijima (1999)] and [Ishijima et al. (2004)].

Versions of the result that, in continuous-time, a two-fund separation result will imply that investors choose fractional Kelly strategies have been shown at different levels of generality in, for example, [Merton (1971, 1973)], [Khanna and Kulldorff (1999)], [Nielsen and Vassalou (2006, 2004)], [Platen (2002)], [Zhao and Ziemba (2003)] and [Christensen and Platen (2005)]. Some general arguments that the GOP will approximate or be identical to the market portfolio are provided in [Platen (2004c, 2005a)]. Chapter 6.3 of [Christensen (2005)] shows that, when the risky asset can be dominated, investors must stay "reasonably close to the GOP" when the market conditions become favorable. However, this is a relatively weak approximation result as it will be made clear. Further results in the case where asset prices are of a more general class are treated in [Platen (2006b)].

In an entirely different literature, the so-called *evolutionary finance* literature, [Blume and Easley (1992)] show that using the GOP will result in market dominance. The conclusion is, however, not stable to more general cases as shown in [Amir *et al.* (2004)], where market prices are determined endogenously and the market is incomplete.

1.3.3. *How Long Does it Take for the GOP to Outperform other Portfolios?*

As the GOP was advocated, not as a particular utility function, but as an alternative to utility theory relying on its ability to outperform other portfolios over time, it is important to document this ability over horizons relevant to actual investors. In this section, it will be assumed that investors are interested in the GOP because they hope it will outperform other competing strategies. This goal may not be a "rational" investment goal from the point of view of expected utility, but it is investigated because it is the predominant reason why the GOP was recommended as an investment strategy, as explained previously.

To achieve a feeling for the time it takes for the GOP to dominate other assets, consider the following illustrative example.

Example 1.10. Assume a situation similar to Example 1.4. This is a two-asset Black–Scholes model with constant parameters. By solving the differential equation, the savings account with a risk-free interest rate of r is given by

$$S^{(0)}(t) = \exp(rt)$$

and solving the SDE, the stock price is given as

$$S^{(1)}(t) = \exp\left((a - \frac{1}{2}\sigma^2)t + \sigma W(t)\right).$$

By Example 1.4, the GOP is given by the process

$$S^{(\delta)}(t) = \exp\left((r + \frac{1}{2}\theta^2)t + \theta W(t)\right),$$

where $\theta = \frac{a-r}{\sigma}$. Some simple calculations imply that the probability

$$P_0(t) \triangleq \mathbb{P}(S^{(\delta)}(t) \geq S^{(0)}(t))$$

of the GOP outperforming the savings account over a period of length t and the probability

$$P_1(t) \triangleq \mathbb{P}(S^{(\delta)}(t) \geq S^{(1)}(t))$$

of the GOP outperforming the stock over a period of length t are given by

$$P_0(t) = N\left(\frac{1}{2}\theta\sqrt{t}\right)$$

and

$$P_1(t) = N\left(\frac{1}{2}|\theta - \sigma|\sqrt{t}\right) .$$

Here, $N(\cdot)$ denotes the cumulative distribution function of the standard Gaussian distribution. Clearly, these probabilities are independent of the short rate. This would remain true even if the short rate was stochastic, as long as the short rate does not influence the market price of risk and volatility of the stock. Moreover, they are increasing in the market price of risk and time horizon. The probabilities converge to one as the time horizon increases to infinity, which is a manifestation of the growth properties of the GOP. Table 1.1 shows the time horizon needed for outperforming the savings account at certain confidence levels. If θ is interpreted as $|\theta - \sigma|$, then the results can be interpreted as the time horizon needed to outperform the stock. For example, if the market price of risk is 0.25 then, over a 105 year period, the GOP will provide a better return than the risk-free asset with a 90% confidence level. This probability is equal to the probability of outperforming a moderately risky stock with a volatility of 50% per year.

Table 1.1. Time horizon needed for outperforming the risk-free asset at certain confidence levels.

Conf. level	$\theta = 0.05$	$\theta = 0.1$	$\theta = 0.25$	$\theta = 0.5$
99%	8659	2165	346	87
95%	4329	1082	173	43
90%	2628	657	105	26

Figure 1.1 illustrates how the outperformance probability depends on the time horizon.

The preliminary conclusion based on these simple results is that the long run may be very long indeed. A Sharpe ratio of 0.5 is a reasonably high value; for example this would be the result of a strategy with an expected

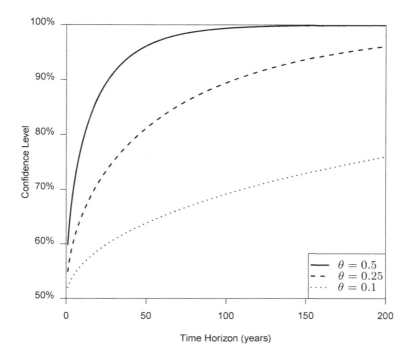

Fig. 1.1. Outperformance likelihood.

excess rate of return above the risk-free rate of 20% and a volatility of 40%. Even with such a strategy, it would take almost 30 years to beat the risk-free bond with a 90% probability.

Similar experiments have been conducted in the literature. For example, [Aucamp (1993)] considers an application of the GOP strategy to the St. Petersburg game and calculates the probability of outperforming a competing strategy. It is analyzed how many games are necessary for the GOP to outperform other strategies at a 95% confidence level. It is shown that this takes quite a number of games. For example, if the alternative is "do nothing", then it takes the growth-optimal betting strategy 87 games. Making the alternative strategy more competitive (i.e. comparing to a conservative betting strategy) makes the number of games required grow very fast. If it takes a long time before the GOP dominates alternative investment strategies, then the argument that one should choose the GOP to maximize the probability of performing better than other portfolios is somewhat weakened. Apart from the discussion of whether this property is interesting or not, it requires an unrealistically long-time horizon to obtain any high level of confidence. For it to be really useful would require the GOP, when calibrated to market data, to outperform, say, an index over a (relatively) short horizon – 10 or 20 years. In this case, given the absence of a clearly-specified utility function it might be useful to consider the GOP strategy. Hence, in order to see how long it will take the GOP to outperform a given alternative strategy, one needs to conduct a further systematic analysis. The analysis needs to be conducted in a more realistic model calibrated to actual market data. There appears to be no available results in this direction in the literature.

Notes

Some papers that include studies of the wealth distribution when applying the GOP include [Hakansson (1971a)], [Gressis *et al.* (1974)], [Michaud (1981)] and [Thorp (2000)]. Somewhat related to this is the study by [Jean (1980)], which relates the GOP to n-th order stochastic dominance. He shows that if a portfolio X exhibits n-th order stochastic dominance against a portfolio Y for any given n, then X needs to have a higher geometric mean than Y.

Example 1.10 is similar to [Rubinstein (1991)], who shows that to be 95% sure of beating an all-cash strategy will require 208 years; to be 95% sure of beating an all-stock strategy will require 4,700 years.

Note that the empirical evidence is mixed, see, for example, the results in [Thorp (1971)], [Hunt (2005a)] and the references in Section 1.5. The existing attempts to apply the GOP seem to have been very successful, but this has the character of "anecdotal evidence" and does not constitute a formal proof that the period required to outperform competing strategies is relatively short.

1.4. The GOP and the Pricing of Financial Assets and Derivatives

The numéraire property of the GOP and Theorem 1.4 has made several authors suggest that it could be used as a convenient pricing tool for derivatives in complete and incomplete markets. Although different motivations and different economic interpretations are possible for this methodology, the essence is very simple. The situation in this section is similar to the general situation described in Section 1.2.2. A set of $d+1$ assets is given as semimartingales and it is assumed that the GOP, $S^{(\delta)}$, exists as a well-defined, non-explosive portfolio process on the interval $[0, T]$. Let us make the following assumption:

Assumption 1.1. For $i \in \{0, \ldots, d\}$, the process

$$\hat{S}^{(i)}(t) \triangleq \frac{S^{(i)}(t)}{S^{(\delta)}(t)}$$

is a local martingale.

Hence, we rule out the cases where the process is a supermartingale but not a local martingale, see Example 1.3. The reason why this is done will become clear shortly. Assumption 1.1 implies that the GOP gives rise to a *martingale density*, in the sense, that for any $S^{(\delta)} \in \underline{\Theta}(S)$, it holds that

$$\hat{S}^{(\delta)}(t) \triangleq \frac{S^{(\delta)}(t)}{S^{(\delta)}(t)} = \frac{S^{(\delta)}(t)}{S^{(0)}(t)} Z(t)$$

is a local martingale, where

$$Z(t) = \frac{S^{(0)}(t)}{S^{(\delta)}(t)} = \hat{S}^{(0)}(t) \ .$$

Therefore, the process $Z(t)$ can, under regularity conditions, be interpreted as the Radon–Nikodym derivative of the usual risk-neutral measure. However, some of these processes may be strict local martingales, not true martingales. In particular, if the GOP-denominated savings account is a true local martingale, then the classical risk-neutral martingale measure will not exist, as will be discussed below.

Definition 1.6. Let H be any $\mathcal{F}_T$-measurable random variable. This random variable is interpreted as the pay-off of some financial asset at time T. Assume that

$$\mathbb{E}\left[\frac{|H|}{S^{(\delta)}(T)}\right] < \infty.$$

The *fair price* process of the pay-off H is then defined as

$$H(t) = S^{(\delta)}(t)\mathbb{E}\left[\frac{H}{S^{(\delta)}(T)}\Big| \mathcal{F}_t\right] . \tag{1.16}$$

The idea is to define the fair price in such a way that the numéraire property of the GOP is undisturbed. In other words, the GOP remains a GOP after the pay-off H is introduced in the market. There are two primary motivations for this methodology. Firstly, the market may not be complete, in which case there may not be a replicating portfolio for the pay-off H. Second, the market may be complete, but there need not exist an equivalent risk-neutral measure, which is usually used for pricing. In the case of complete markets which have an equivalent risk-neutral measure, the fair-pricing concept is equivalent to pricing using the standard method.

Lemma 1.2. *Suppose the market has an equivalent martingale measure, that is, a probability measure Q such that $P \sim Q$, and discounted asset prices are Q-local martingales. Then, the risk-neutral price given by*

$$\tilde{H}(t) = S^{(0)}(t)\mathbb{E}^Q\left[\frac{H}{S^{(0)}(T)}\Big|\mathcal{F}_t\right]$$

is identical to the fair price, i.e. $H(t) = \tilde{H}(t)$ almost surely, for all $t \in [0, T]$.

The following example illustrates why fair-pricing is restricted as suggested by Assumption 1.1.

Example 1.11. (Example 1.3 continued) Recall that the market is given such that the first asset is risk-free, $S^{(0)}(t) = 1$, $t \in \{0, T\}$ and the second asset has a log-normal distribution $\log(S^{(1)}(T)) \sim \mathcal{N}(\mu, \sigma^2)$ and $S^{(1)}(0) = 1$.

Suppose that $\hat{S}^{(0)}(t)$ is a strict supermartingale. What happens if the fair-pricing concept is applied to a zero-coupon bond? The price of the zero-coupon bond in the market is simply $S^{(0)}(0) = 1$. The fair price on the other hand is

$$S^{(1)}(0)\mathbb{E}\left[\frac{1}{S^{(1)}(T)}\right] < 1.$$

Hence, introducing a fairly-priced, zero-coupon bond in this market produces an arbitrage opportunity. More generally, this problem will occur in all cases where some primary assets denoted in units of the GOP are strict supermartingales, and not local martingales.

Below we consider the remaining cases in turn. In the incomplete market case, we show how the fair price defined above is related to other pricing methodologies. Then we consider markets without a risk-neutral measure and discuss how and why the GOP can be used in this case.

1.4.1. *Incomplete Markets*

Fair pricing as defined above was initially suggested as a method for pricing derivatives in incomplete markets, see [Bajeux-Besnaino and Portait (1997a)] and the sources cited in the notes. In this subsection, markets are assumed to be incomplete, but, to keep things separate, it is assumed that the set of martingale measures is non-empty. In particular, the process $\hat{S}^{(0)}$ is assumed to be a true martingale. When markets are incomplete and there is no portfolio which replicates the pay-off, H, arbitrage theory is silent on how to price this pay-off. From the seminal work of [Harrison and Pliska (1981)] it is well-known that this corresponds to the case of an infinite number of candidate martingale measures. Any of these measures will price financial assets in accordance with no-arbitrage and there is no *a-priori* reason for choosing one over the other. In particular, using no arbitrage considerations does not suggest that one might use the martingale measure Q defined by $\frac{dQ}{dP} = \hat{S}^{(0)}(T)$, which is the measure induced by applying the GOP. One might assume that investors maximized the growth rate of their investments. Then it could be argued that a "reasonable" price of the pay-off, H, should be such that the maximum growth rate obtainable from trading the derivative and the existing assets should not be higher than trading the existing assets alone. Otherwise, the derivative would be in positive net demand, as investors applied it to obtain a higher growth rate. It can be shown that the only pricing rule which satisfies this property is the fair-pricing rule. Of course, whether or not growth rates are interesting to investors has been a controversial issue. Indeed, as outlined in the previous sections, the growth rate is only indirectly relevant to an investor with logarithmic utility. The argument that the maximal growth rate should not increase after the introduction of the derivative is generally not backed by an equilibrium argument, except for the case where the representative investor is assumed to have logarithmic utility. Although there may be no strong theoretical argument behind the selection of the GOP as the pricing operator in an incomplete market, its application is fully consistent with arbitrage-free pricing. Consequently, it is useful to compare this method to a few of the pricing alternatives presented in the literature.

1.4.1.1. *Utility-Based Pricing*

This approach to pricing assumes agents to be endowed with some utility function U. The *utility indifference price* at time t of k units of the pay-off

H, is then defined as the price $p_H(k, t)$ such that

$$\sup_{S^{(\delta)}(T), S^{(\delta)}(t)=x-p_H(k,t)} \mathbb{E}\left[U(S^{(\delta)}(T) + kH)\right]$$

$$= \sup_{S^{(\delta)}(T), S^{(\delta)}(t)=x} \mathbb{E}\left[U(S^{(\delta)}(T))\right].$$

This price generally depends on k, i.e., on the number of units of the pay-off, in a non-linear fashion, due to the concavity of U. Supposing that the function $p_H(k, t)$ is smooth, one may define the *marginal price* as the limit

$$p_H(t) = \lim_{k \to 0} \frac{p_H(k, t)}{k},$$

which is the utility indifference price for obtaining a marginal unit of the pay-off, when the investor has none to begin with. If one uses logarithmic utility, then the marginal indifference price is equal to the fair price, i.e. $p_H(t) = H(t)$. Of course, any reasonable utility function could be used to define a marginal price, the logarithm is only a special case.

1.4.1.2. *The Minimal Martingale Measure*

This is a particular choice of measure, which is often selected because it "disturbs" the model as little as possible. This is to be understood in the sense that a process which is independent of traded assets will have the same distribution under the minimal martingale measure as under the original measure. Assume the semimartingale, S, is special such that it has locally-integrable jumps and consequently has the unique decomposition

$$S(t) = S(0) + A(t) + M(t),$$

where $A(0) = M(0) = 0$, A is predictable and of finite variation, and M is a local martingale. In this case, one may write

$$dS(t) = \lambda(t)d\langle M\rangle_t + dM(t),$$

where λ is the market price of risk process and $\langle M\rangle$ is the predictable projection of the quadratic variation of M. The minimal martingale measure, if it exists, is defined as the solution to the SDE

$$dZ(t) = -\lambda(t)Z(t)dM(t).$$

In financial terms, the minimal martingale measure puts the market price of any unspanned risk, that is, risk factors that cannot be hedged by trading in the market, equal to zero. In the general case, Z may not be a martingale, and it may become negative. In such cases the minimal martingale measure is not a true probability measure. If S is continuous, then using the minimal martingale measure provides the same prices as the fair-pricing concept. In the general case, when asset prices may exhibit jumps, the two methods for pricing assets are generally different.

1.4.1.3. *Good-Deal Bounds*

Some authors have proposed pricing claims by defining a bound on the market prices of risk that can exist in the market. Choosing a martingale measure in an incomplete market amounts to the choice of a specific market price of risk. As mentioned, the minimal martingale measure is obtained by setting the market price of risk of non-traded risk factors equal to zero. For this reason, the price derived from the minimal martingale measure always lies within the good-deal bounds. Given the assumption that the set of prices within the good-deal bound is non-empty. It follows that the fair price lies within the good-deal bound in the case of continuous asset prices. In the general case, the fair price need not lie within a particular good deal bound.

Another application of fair-pricing is found in the Benchmark approach. However, here the motivation was somewhat different as it will be described below.

Notes

The idea of using the GOP for pricing purposes is stated explicitly for the first time in the papers [Bajeux-Besnaino and Portait (1997a,b)] and further argued in [Aurell *et al.* (2000a,b)]. In the latter case, the arguments for using the GOP seem to be subject to the criticism raised by Samuelson, but the method, as such, is not inconsistent with no arbitrage. Utility-based pricing is reviewed in [Davis (1997)] and [Henderson and Hobson (2008)]. The minimal martingale measure is discussed in, for example [Schweizer (1995)], and the relationship with the GOP is discussed in [Becherer (2001)] and [Christensen and Larsen (2007)]. Good-deal bounds were introduced by [Cochrane and Saá-Requejo (2000)] and extended to a general setting in [Björk and Slinko (2006)]. The later has a discussion of the relationship to the minimal martingale measure.

1.4.2. *A World Without a Risk-Neutral Measure*

In this section a complete market is considered. To keep matters as simple as possible, assume there is a risk-free savings account where the short rate, r, is assumed to be constant. Hence,

$$dS^{(0)}(t) = rS^{(0)}(t)dt.$$

There is only one risky asset given by the stochastic differential equation

$$dS^{(1)}(t) = S^{(1)}(t)\left(a(t)dt + b(t)dW(t)\right),$$

where W is a standard one-dimensional Wiener process. It is assumed that a and b are strictly-positive processes such that the solution $S^{(1)}$ is unique and well-defined, but no other assumptions are made. The parameter processes a and b can be general stochastic processes. The market price of risk θ is then well-defined as $\theta(t) = \frac{a(t)-r(t)}{b(t)}$ and consequently this may also be a stochastic process. The usual approach when pricing options and other derivatives is to define the stochastic exponential

$$\Lambda(t) = \exp\left(-\frac{1}{2}\int_0^t \theta^2(s)ds - \int_0^t \theta(s)dW(s)\right).$$

If $\Lambda(t)$ is a martingale, then the Girsanov theorem implies the existence of a measure Q such that

$$\tilde{W}(t) \triangleq W(t) - \int_0^t \theta(s)ds$$

is a standard Wiener process under the measure Q.

However, it is well-known that Λ need not be a martingale. This is remarked in most textbooks, see, for example, [Karatzas and Shreve (1988)] or [Revuz and Yor (1991)]. The latter contains examples from the class of Bessel processes.

By the Itô formula, Λ satisfies the stochastic differential equation

$$d\Lambda(t) = -\theta(t)\Lambda(t)dW(t)$$

and so is a local martingale, see [Karatzas and Shreve (1988)]. As mentioned in Example 1.9, some additional conditions are required to ensure Λ is a martingale. The question here is: what happens if the process $\Lambda(t)$ is not a martingale? The answer is given in the theorem below:

Theorem 1.9. *Suppose the process* $\Lambda(t)$ *is not a true martingale. Then*

(1) If there is a stopping time $\tau \leq T$, *such that* $\mathbb{P}(\int_0^\tau \theta^2(s)ds = \infty) > 0$,
then there is no equivalent martingale measure for the market under any numéraire and the GOP explodes. An attempt to apply risk-neutral pricing or fair-pricing will result in Arrow–Debreu prices that are zero for events with positive probability.

(2) If $\int_0^T \theta^2(s)ds < \infty$ *almost surely, then the GOP is well-defined and the original measure* P *is an equivalent martingale measure when using the GOP as numéraire.*

(3) If $\int_0^T \theta^2(s)ds < \infty$ *almost surely, then* $\Lambda(t)$ *is a strict supermartingale and there is no risk-neutral measure when using the risk-free asset as a numéraire. Moreover, the risk-free asset can be outperformed over some interval* $[0, \tau] \subseteq [0, T]$.

(4) The fair price is the price of the cheapest portfolio that replicates the given pay-off.

The theorem shows that fair-pricing is well-defined in cases where risk-neutral pricing is not. Although presented here in a very special case, the result is, in fact, true in a very general setting. The result may look puzzling at first because usually the existence of a risk-neutral measure is associated with the absence of arbitrage. However, continuous-time models may contain certain types of "arbitrage" arising from the ability to conduct an infinite number of trades. A prime example is the so-called doubling strategy, which involves doubling the investment until the time when a favorable event happens and the investor realizes a profit. Such "arbitrage" strategies are easily ruled out as being inadmissible by Definition 1.4 because they generally require an infinite debt capacity. Hence, they are not arbitrage strategies in the sense of Definition 1.3. However, imagine a not-so-smart investor, who tries to do the opposite thing. He may end up losing money, with certainty, by applying a so-called "suicide strategy", which is a strategy that costs money but results in zero terminal wealth. A suicide strategy could, for example, be a short position in the doubling strategy (if it where admissible). Suicide strategies exist *whenever asset prices are unbounded* and they need not be inadmissible. Hence, they exist in, for example, the Black–Scholes model and other popular models of finance. If a primary asset has a built-in suicide strategy, the asset can be outperformed by a replicating portfolio without the suicide strategy. This suggests the existence of an arbitrage opportunity, but that is not the case. If an investor attempts to sell the asset and buy a (cheaper) replicating portfolio, the resulting strategy is not necessarily admissible. Indeed, this strategy may suffer large, temporary losses before maturity, at which point, of course, it becomes strictly positive. It is important to note that whether or not the temporary losses of the portfolio are bounded is strictly dependent on the numéraire. This insight was developed by [Delbaen and Schachermayer (1995b)], who showed that the arbitrage strategy under consideration is *lower-bounded* under some numéraire, if and only if that numéraire can be outperformed. Given the existence of a market price of risk, the "arbitrage" strategy is never strictly positive at all times before maturity. If this was the case, then any investor could take an unlimited position in this arbitrage and the GOP would no longer be a well-defined object. The important difference between having a lower-bounded and an unbounded arbitrage strategy is that, if the strategy is lower-bounded, then, by the fundamental theorem of asset pricing, there can be no equivalent martingale measure. In particular, if the risk-free asset contains a built-in suicide strategy, then there cannot be an EMM when the risk-free asset is applied as numéraire. On the other hand, a different numéraire may still work allowing for consistent pricing of derivatives.

Hence, if one chooses a numéraire which happens to contain a built-in suicide strategy, then one cannot have an equivalent martingale measure. This suggests that one need only take care that the right numéraire is chosen, in order to have a consistent theory for pricing and hedging. Indeed, the GOP is such a numéraire, and it works even when the standard numéraire – the risk-free savings account – can be outperformed. Moreover, the existence of a GOP is completely numéraire-independent. In other words, the existence of a GOP is the acid test of any model that is to be useful when pricing derivatives.

Obviously, the usefulness of the described approach relies on two things. Firstly, there is only a theoretical advantage in the cases where risk-neutral pricing fails, otherwise the two approaches are completely equivalent, and fair-pricing is merely a special case of the change of numéraire technique.

Remark 1.1. In principle, many other numéraires could be used instead of the GOP, as long as they do not contain suicide strategies. For example, the portfolio choice of a utility-maximizing investor with increasing utility function will never contain a suicide strategy. However, for theoretical reasons, the GOP is more convenient. Establishing the existence of a portfolio which maximizes the utility of a given investor is non-trivial. Moreover, the existence of a solution may depend on the numéraire selected, whereas the existence of the GOP does not.

In practice, usefulness requires documentation that the risk-free asset can be outperformed, or, equivalently, that the risk-free asset denominated in units of the GOP is a strict local martingale. This is quite a difficult task. The question of whether the risk-free asset can be outperformed is subject to the so-called peso problem: only one sample path is ever observed, so it is quite hard to argue that a portfolio exists which can outperform the savings account *almost surely*. At the end of the day, "almost surely" means with probability one, not probability 99.999%. From the earlier discussion, it is known that the GOP will outperform the risk-less asset sooner or later, so over a very long time horizon, the probability that the risk-free asset is outperformed is rather high and it is more than likely that even if one were to have (or construct) a number of observations, they would all suggest that the risk-free asset could be outperformed. A better, and certainly more feasible, approach, if one were to document the usefulness of the fair-pricing approach is to show that the predictions of models in which the savings account can be outperformed are in line with empirical observations. Some arguments have started to appear, for example in the literature on continuous-time "bubbles", cited in the notes.

Notes

In a longer sequence of papers the fair-pricing concept was explored as part of the so-called *benchmark approach*, advocated by Eckhard Platen and a number of co-authors, see, for example, [Heath and Platen (2002a,b,c, 2003)], [Miller and Platen (2005)], [Platen (2001, 2002, 2004a,b,c,d, 2005a)], [Christensen and Platen (2005)] and [Platen and West (2005)].

The proof of Theorem 1.9 is found in this literature, see in particular [Christensen and Larsen (2007)]. Some calibrations, which indicate that models without an EMM could be realistic, are presented in [Platen (2004c)] and [Fergusson and Platen (2006)].

Related to this approach is the recent literature on continuous-asset price bubbles. Bubbles are said to exist whenever an asset contains a built-in suicide strategy, because, in this case, the current price of the asset is higher than the price of a replicating portfolio. References are [Loewenstein and Willard (2000a,b)], [Cassese (2005)] and [Cox and Hobson (2005)]. In this literature, it is shown that bubbles can be compatible with equilibrium and that they are in line with observed empirical observations. To some extent, they lend further support to the relevance of fair-pricing.

1.5. Empirical Studies of the GOP

Empirical studies of the GOP are relatively limited in number. Many date back to the seventies and deal with comparisons with the mean-variance model. Here, the empirical studies will be separated into two major groups, according to broad questions:

- How is the GOP composed? Issues that belong to this group include what mix of assets constitutes the GOP and, in particular, whether the GOP equals the market portfolio or any other diversified portfolio.
- How does the GOP perform? Given an estimate of the GOP, it is of some practical importance to document its value as an investment strategy.

The conclusions within these areas are reasonably consistent across the literature and the main ones are

- It is statistically difficult to separate the GOP from other portfolios – this conclusion appears in all studies known to the author. It appears that the GOP is riskier than the mean-variance tangency portfolio and the market portfolio, but the hypothesis that the GOP is the market portfolio cannot be formally rejected. It may be well-approximated by a levered position in the market. This is consistent with different estimates of the risk aversion coefficient, γ, of a power-utility investor, which different authors have estimated to be much higher than one (corresponding to a log-investor). A problem in most studies is the lack of statistical significance and it is hard to find significant proof of the composition. Often, running some optimization program will imply a GOP that only invests in a smaller subset of available assets.
- The studies that use the GOP for investment purposes generally conclude that, although it may be subject to large short-term fluctuations, growth maximization performs rather well even on time horizons which are not excessively long. Hence, although the GOP does not maximize expected utility for non-logarithmic investors, history shows that portfolio managers using the GOP strategy can become rather successful. However, the cases where the GOP is applied for investment purposes are of a somewhat anecdotal nature. Consequently, the focus of this section will be the first question.

Notes

For some interesting reading on the actual performance of growth opti-mal portfolios in various connections, see [Thorp (1971, 2000)], [Grauer and Hakansson (1985)], [Hunt (2005b,a)] and [Ziemba (2005)]. Edward Thorp constructed a hedge-fund, PNP, which successfully applied the GOP strat-egy to exploit prices in the market out of line with mathematical models, see in particular [Poundstone (2005)]. The reader is referred to the quoted papers in the following subsection, since most of these have results on the performance of the GOP as well. There seem to be very few formal studies which consider the performance of growth-optimal strategies.

1.5.1. *Composition of the GOP*

1.5.1.1. *Discrete-Time Models*

The method used for empirical testing is replicated, at least in princi-ple, by several authors and so is explained briefly. Assume a discrete-time case, as described in Section 1.2.1. Hence, the market is given as $S = (S^{(0)}(t), S^{(1)}(t), \ldots, S^{(d)}(t))$, with the return between time t and $t+1$ for asset i denoted by $R^i(t)$ as usual. Recall from the myopic properties of the GOP that the GOP strategy can be found by maximizing the expected growth rate between t and $t+1$,

$$\sup_{\delta} \ \mathbb{E}_t \left[\log \left(\frac{S^{(\delta)}(t+1)}{S^{(\delta)}(t)} \right) \right],$$

for each $t \in \{0, \ldots T-1\}$. From Equation (1.5), the first-order conditions for this problem are

$$\mathbb{E}_{t-1} \left[\frac{1 + R^i(t)}{1 + R^{\underline{\delta}}(t)} \right] = 1 \qquad (1.17)$$

for all $i \in \{0, \ldots, d\}$. The first-order conditions provide a testable impli-cation. A test that some given portfolio is the GOP can be carried out by forming samples of the quantity

$$Z^i(t) \triangleq \frac{1 + R^i(t)}{1 + R^{\underline{\delta}}(t)},$$

where $1 + R^{\underline{\delta}}(t)$ is the return of candidate GOP portfolio. The test consists of checking whether

$$\bar{Z} \triangleq \frac{1}{T} \sum_{t=1}^{T} Z_t^i$$

is statistically different across assets. However, note that this requires the first-order conditions to be satisfied. In theory, GOP-deflated assets are supermartingales – in particular, they may, in discrete-time, be supermartingales that are not martingales, see Example 1.3. The assumption implicit in the first-order condition above is that optimal GOP fractions are assumed in an inner point. A theoretically more correct approach would be to test whether these quantities, on average, are below one. As the variance of returns may differ across assets and independence is unrealistic, an applied approach is to use the Hotelling T^2 statistic to test this hypothesis. This is a generalization of the student-t distribution. To be a valid test, this actually requires normality from the underlying variables, which is clearly unreasonable, since, if returns would have support on the entire real axis, a growth optimizer would seek the risk-free asset. The general conclusion from this approach is that it cannot reject the hypothesis that the GOP equals the market portfolio.

Because this approach is somewhat dubious, alternatives have been suggested. An entirely different way of solving the problem is to find the GOP in the market by making more specific distributional assumptions, and calculating the GOP *ex ante* and studying its properties. This allows a comparison between the theoretically-calculated GOP and the market portfolio. The evidence is somewhat mixed. [Fama and Macbeth (1974)] compare the mean-variance efficient tangent portfolio to the GOP. Perhaps the most important conclusion of this exercise is that, although the β of the historical GOP is large and deviates from one, the growth rate of this portfolio cannot be statistically separated from that of the market portfolio. This is possibly related to the fact that [Fama and Macbeth (1974)] construct the time series of growth rates from relatively short periods and, therefore, the size of growth rates is reasonably small compared to the sample variance which, in turn, implies small t-stats. Still, it suggests that the GOP could be more highly-levered than the tangency portfolio. This does not imply, of course, that the GOP is different from the market. This would be postulating a beta of one to be the beta of the market portfolio and it requires one to believe that the CAPM holds.

Although the cited study finds it difficult to reject the proposition that the market portfolio is a proxy for the GOP, it suggests that the GOP can be more risky than this portfolio. Note that the market portfolio itself has to be proxied. Usually this is done by taking a large index such as S&P 500 as a proxy for the market portfolio. Whether this approximation is reasonable is debatable. Indeed, this result is verified in most available studies. An exception is [Long (1990)], who examines different proxies for the GOP. The suggested proxies are examined using the first-order conditions as described above. Although the results of formal statistic tests are absent, the intuition is in line with earlier empirical research. In the article, three proxies are examined:

(1) A market portfolio proxy.
(2) A levered position in the market portfolio.
(3) A quasi-maximum-likelihood estimate.

The study concludes that using a quasi-maximum-likelihood estimate of the GOP exhibits superior performance. However, using a market portfolio proxy as numéraire will yield a time series of numéraire-adjusted returns which has a mean close to zero. A levered position in the market portfolio, on the other hand, will increase the variance of the numéraire adjusted returns and seems to be the worst option.

A general conclusion, when calibrating the market data to a CAPM type model, is that the implied relative risk aversion for a representative agent is (much) higher than one, one being the relative risk aversion of an agent with logarithmic utility. This somehow supports the conclusion that the GOP is more risky than the market portfolio.

A few other studies indicate that the GOP could be a rather narrow portfolio, selecting only a few stocks. A study which deals more specifically with the question of what assets to include in the GOP was conducted by [Grauer (1981)]. Assuming that returns on assets follow a (discrete) approximate normal distribution, he compares the mix of assets in a mean-variance efficient portfolio and a GOP, with limits on short sales. Out of twenty stocks, the GOP and the mean-variance tangency portfolio both appeared to be very *undiversified* – the typical number of stocks picked by the GOP strategy was three. Similar experimental results can be found in Chapters 2 and 4 of this volume. Furthermore, there appeared to be a significant difference between the composition of a mean-variance efficient portfolio and the GOP. In a footnote, Grauer suggests that this may be due to the small number of states (which makes hedging simpler) in the distribution of returns. This does not explain the phenomena for the tangency portfolio, which, if CAPM holds, should have the same composition as the market portfolio. It suggested that the lack of diversification is caused by the imposed short-sale constraint. Although the reason for this explanation is unclear, it is shown that, if the short-sale constraint is lifted, then the GOP model takes a position in all 20 stocks. In [Grauer and Hakansson

(1985)] a simple model for returns is assumed and the investor's problem is then solved using non-linear programming. It appears that the GOP is well-approximated by this method and it yields a significantly higher mean geometric return than other strategies investigated in the sample. Analyzing the composition of the GOP provides a somewhat mixed picture of diversification; before 1940, the GOP consists of a highly-levered position in Government and Corporate bonds, but only a few stocks. A switch then occurs towards a highly-levered position in stocks until the late sixties, at which point the GOP almost leaves the stock market and turns to the risk-free asset to become a quite conservative strategy in the last period of the sample, which ends in 1982. This last period of conservatism may be due to estimation problems and it is remarked by the authors that, by analyzing the *ex post* returns, it appears that the GOP is *too* conservative. Still, the article is not able to support the claim that the GOP, in general, is a well-diversified portfolio.

1.5.1.2. *Continuous Time Models*

Only a very few studies have been made in continuous-time. An example is [Hunt (2005b)], who uses a geometric Brownian motion with one driving factor as the model for stock prices. This implies that stocks are perfectly correlated across assets and log returns are normally-distributed. Despite the fact that such a model is rejected by the data, a GOP is constructed, and its properties are investigated. The GOP strategy formed in this setting also consists of only a few stocks, but imposing a short-sale constraint *increases* the level of diversification in GOP strategy, contrary to the result mentioned above. The study is subject to high parameter uncertainty, and the assumption of one driving factor implies that the GOP strategy is far from unique; in theory, it can be formed from any two distinct assets. For this reason, the conclusions about the composition of the GOP might be heavily influenced by the choice of model.

It appears that, to answer the question of what mix of assets are required to form the GOP, new studies will have to be made. In particular, obtaining closer approximation of real stock dynamics is warranted. This could potentially include jumps and should at least have several underlying uncertainty factors driving returns. The overall problem so far seems to have been a lack of statistical power, but certainly having a realistic underlying model seems to be a natural first step. Furthermore, the standard test in Equation (1.5) may be insufficient if the dynamics of GOP deflated assets will be that of a true supermartingale. The test may lead to an acceptance of the hypothesis that a given portfolio is growth-optimal, when the true portfolio is, in fact, more complex. Hence, tests based on the first-order condition should, in principle, be one-sided.

Notes

Possibly the first study to contain an extensive empirical study of the GOP was [Roll (1973)]. Both [Roll (1973)] and [Fama and Macbeth (1974)] suggest that the market portfolio should approximate the GOP, both use an index as a GOP candidate, and both are unable to reject the conclusion that the GOP is well-approximated by the market portfolio. [Roll (1973)] use the S&P 500, whereas [Fama and Macbeth (1974)] use a simple average of returns on common stocks listed on New York Stock Exchange (NYSE). Using the first-order condition as a test of growth-optimality is also done by [Long (1990)] and [Hentschel and Long (2004)]. [Bicksler and Thorp (1973)] assume two different distributions, calculate the implied GOP based on different amounts of leverage and find it to be highly-levered. Second, many growth *suboptimal* portfolios are impossible to separate from the true GOP in practice. [Rotando and Thorp (1992)] calibrate S&P 500 data to a truncated normal distribution and calculate the GOP formed by the index and risk-less borrowing. This results in a levered position in the index of about 117%. [Pulley (1983)] also reaches the conclusion that the GOP is not a very diversified portfolio. However, in Pulley's study, the composition of the GOP and mean-variance based approximations are very similar, see [Pulley (1983)][Table 2]. For general results suggesting the market portfolio to be the result of a representative agent with high risk aversion, see, for example, the econometric literature related to the equity premium puzzle of [Mehra and Prescott (1985)]. Some experimental evidence is presented in [Gordon *et al.* (1972)], showing that as individuals become more wealthy, their investment strategy would approximate that of the GOP. [Maier *et al.* (1977a)] conduct simulation studies to investigate the composition of the GOP and the performance of various proxies.

1.6. Conclusion

The GOP has fascinated academics and practitioners for decades. Despite the arguments made by respected economists that the growth properties of the GOP are irrelevant as a theoretical foundation for portfolio choice, it appears that it is still viewed as a practically-applicable criterion for investment decisions. In this debate it was emphasized that the utility paradigm in comparison suffers from being somewhat more abstract. The arguments that support the choice of the GOP are based on very specific growth properties, and even though the GOP is the choice of a logarithmic investor, this interpretation is often just viewed as coincidental. The fact that over time the GOP will outperform other strategies is an intuitively appealing property, since when the time comes to liquidate the portfolio it only matters how much money it is worth. Still, some misunderstandings seem to persist in this area, and the fallacy pointed out by Samuelson probably should be studied more carefully by would-be applicants of this strategy, before they make their decision. Moreover, the dominance of the GOP may require some patience. Studies show that it will take many years before the probability becomes high that the GOP will do better than even the risk-free asset.

In recent years, it is, in particular, the numéraire property of the GOP which is being researched. This property relates the GOP to pricing kernels and, hence, makes it applicable for pricing derivatives. Hence, it appears that the GOP may have a role to play as a tool for asset and derivative pricing. The practical applicability and usefulness still need to be validated empirically, in particular the problem of finding a well-working GOP proxy needs attention. This appears to be an area for further research in the years to come.

References

Aase, K. K. (1984). Optimum portfolio diversification in a general continuous-time model, *Stochastic Processes and their Applications* **18**, 1, pp. 81–98.

Aase, K. K. (1986). Ruin problems and myopic portfolio optimization in continuous trading, *Stochastic Processes and their Applications* **21**, 2, pp. 213–227.

Aase, K. K. (1988). Contingent claim valuation when the security price is a combination of an Itô process and a random point process, *Stochastic Processes and their Applications* **28**, 2, pp. 185–220.

Aase, K. K. (2001). On the St. Petersburg paradox, *Scandinavian Actuarial Journal.* **3**, 1, pp. 69–78.

Aase, K. K. and Oksendal, B. (1988). Admissible investment strategies in continuous trading, *Stochastic Processes and their Applications* **30**, 2, pp. 291–301.

Algoet, P. and Cover, T. (1988). Asymptotic optimality and asymptotic equipartition properties of log-optimum investment, *The Annals of Probability* **16**, 2, pp. 876–898.

Amir, R., Evstigneev, I., Hens, T. and Schenk-Hoppé, K. R. (2004). Market selection and survival of investment strategies, *Journal of Mathematical Economics* **41**, 1, pp. 105–122.

Ammendinger, J., Imkeller, P. and Schweizer, M. (1998). Additional logarithmic utility of an insider, *Stochastic Processes and their Applications* **75**, 2, pp. 263–286.

Aucamp, D. (1977). An investment strategy with overshoot rebates which minimizes the time to attain a specified goal, *Management Science* **23**, 11, pp. 1234–1241.

Aucamp, D. (1993). On the extensive number of plays to achieve superior performance with the geometric mean strategy, *Management Science* **39**, 9, pp. 1163–1172.

Aurell, E., Baviera, R., Hammarlid, O., Serva, M. and Vulpiani, A. (2000a). Gambling growth optimal investment and pricing of derivatives, *Physica A* **280**, pp. 505–521.

Aurell, E., Baviera, R., Hammarlid, O., Serva, M. and Vulpiani, A. (2000b). A general methodology to price and hedge derivatives in incomplete markets, *International Journal of Theoretical and Applied Finance* **3**, 1, pp. 1–24.

Aurell, E. and Muratore-Ginanneschi, P. (2004). Growth-optimal strategies with quadratic friction over finite-time investment horizons, *International Journal of Theoretical and Applied Finance* **7**, pp. 645–657.

Bajeux-Besnaino, I. and Portait, R. (1997a). The numéraire portfolio: A new perspective on financial theory, *The European Journal of Finance* **3**, 4, pp. 291–309.

Bajeux-Besnaino, I. and Portait, R. (1997b). Pricing contingent claims in incomplete markets using the numéraire portfolio, Working Paper. George Washington University.

Becherer, D. (2001). The numéraire portfolio for unbounded semi-martingales, *Finance and Stochastics* **5**, 3, pp. 327–341.

Bell, R. and Cover, T. (1980). Competitive optimality of logarithmic investment, *Mathematics of Operations Research* **5**, 2, pp. 161–166.

Bell, R. and Cover, T. (1988). Game-theoretic optimal portfolios, *Management Science* **34**, 6, pp. 724–733.

Bellman, R. and Kalaba, R. (1957). Dynamic programming and statistical communication theory, *Proceedings of the National Academy of Sciences of the United States of America* **43**, 8, pp. 749–751.

Benartzi, S. and Thaler, R. H. (1999). Risk aversion or myopia? choices in repeated gambles and retirement investments, *Management Science* **45**, 3, pp. 364–381.

Bernoulli, D. (1954). Exposition of a new theory on the measurement of risk, *Econometrica* **22**, 1, pp. 23–36.

Bickel, S. H. (1969). Minimum variance and optimal asymptotic portfolios, *Management Science* **16**, 3, pp. 221–226.

Bicksler, J. L. and Thorp, E. (1973). The capital growth model: An empirical investigation, *Journal of Financial and Quantitative Analysis* **8**, 2, pp. 273–287.

Björk, T. and Slinko, I. (2006). Towards a general theory of good deal bounds, *Review of Finance* **10**, pp. 221–260.

Blazenko, G., MacLean, L. and Ziemba, W. T. (1992). Growth versus security in dynamic investment analysis, *Management Science* **38**, 11, pp. 1562–1585.

Blume, L. and Easley, D. (1992). Evolution and market behavior, *Journal of Economic Theory* **58**, 1, pp. 9–40.

Bossaerts, P. (2002). *Paradox of Asset Pricing* (Princeton University Press).

Breiman, L. (1960). Investment policies for expanding businesses optimal in a long-run sense, *Naval Research Logistics Quarterly* **7**, 4, pp. 647–651.

Breiman, L. (1961). Optimal gambling systems for favorable games, *4th Berkeley Symposium on Probability and Statistics* **1**, pp. 65–78.

Brock, W. A. (1970). An axiomatic basis for the Ramsey–Weiszacker overtaking criterion, *Econometrica* **38**, 6, pp. 927–929.

Brouwer, P. D. and den Spiegel, F. V. (2001). The fallacy of large numbers revisited: The construction of a utility function that leads to the acceptance of two games while one is rejected, *Journal of Asset Management* **1**, 3, pp. 257–266.

Browne, S. (1999). Optimal growth in continuous-time with credit risk, *Probability in Engineering and Information Science* **13**, 2, pp. 129–145.

Budd, A. and Litzenberger, R. (1971). A note on geometric-mean portfolio selection and the market prices of equity, *Journal of Financial and Quantitative Analysis* **6**, 5, pp. 1277–1282.

Bühlmann, H. and Platen, E. (2003). A discrete-time benchmark approach to insurance and finance, *ASTIN Bulletin* **33**, 2, pp. 153–172.

Cassesse, G. (2005). A note on asset bubbles in continuous time, *International Journal of Theoretical and Applied Finance* **8**, 4, pp. 523–537.

Christensen, M. M. (2005). *A Thesis on the Growth Optimal Portfolio and the concept of Arbitrage Pricing*, Ph.D. thesis, University of Southern Denmark.

Christensen, M. M. and Larsen, K. (2007). No arbitrage and the growth optimal portfolio, *Stochastic Analysis and Applications* **25**, 5, pp. 255–280.

Christensen, M. M. and Platen, E. (2005). A general benchmark model for stochastic jumps, *Stochastic Analysis and Applications* **23**, 5, pp. 1017–1044.

Cochrane, J. H. and Saá-Requejo, J. (2000). Beyond arbitrage: Good-deal asset price bounds in incomplete markets, *Journal of Political Economy* **108**, 1, pp. 79–119.

Cover, T. (1984). An algorithm for maximizing expected log investment return, *IEEE Transactions on Information Theory* **30**, 2, pp. 369–373.

Cover, T. (1991). Universal portfolios, *Mathematical Finance* **1**, 1, pp. 1–29.

Cover, T. and Iyengar, G. (2000). Growth-optimal investments in horse-race markets with costs, *IEEE Transactions on Information Theory* **46**, 7, pp. 2675–2683.

Cox, A. and Hobson, D. (2005). Local martingales, bubbles and option prices, *Finance and Stochastics* **9**, 5, pp. 477–492.

Cvitanić, J. and Karatzas, I. (1992). Convex duality in constrained portfolio optimization, *The Annals of Applied Probability* **2**, 4, pp. 767–818.

Davis, M. H. A. (1997). Option pricing in incomplete markets. in M. Dempster and S. Pliska (eds.), *Mathematics of Derivative Securities* (Cambridge University Press), pp. 216–227.

Delbaen, F. and Schachermayer, W. (1994). A general version of the fundamental theorem of asset pricing, *Mathematische Annalen* **300**, 3, pp. 463–520.

Delbaen, F. and Schachermayer, W. (1995a). Arbitrage posibilities in Bessel processes and their relation to local martingales, *Probability Theory and Related Fields* **102**, 3, pp. 357–366.

Delbaen, F. and Schachermayer, W. (1995b). The no-arbitrage property under a change of numéraire, *Stochastics and Stochastic Reports* **53**, pp. 213–226.

Delbaen, F. and Schachermayer, W. (1998). The fundamental theorem of asset pricing for unbounded stochastic processes, *Mathematische Annalen* **312**, 2, pp. 215–250.

Elton, E. J. and Gruber, M. J. (1974). On the maximization of the geometric mean with log-normal return distribution, *Management Science* **21**, 4, pp. 483–488.

Elworthy, K., Li, X. and Yor, M. (1999). The importance of strictly local martingales; applications to radial Ornstein-Uhlenbeck processes, *Probab. Theory Related Fields* **115**, 3, pp. 325–355.

Ethier, S. N. (2004). The Kelly system maximizes median wealth, *Journal of Applied Probability* **41**, 4, pp. 1230–1236.

Fama, E. F. and Macbeth, J. D. (1974). Long-term growth in a short-term market, *The Journal of Finance* **29**, 3, pp. 857–885.

Fergusson, K. and Platen, E. (2006). On the distributional characterization of a world stock index, *Applied Mathematical Finance* **13**, 1, pp. 19–38.

Gerard, B., Santis, G. D. and Ortu, F. (2000). Generalized numéraire portfolios, *UCLA Working Papers in Finance*.

Goldman, M. B. (1974). A negative report on the "near optimality" of the max-expected-log policy as applied to bounded utilities for long-lived programs, *Journal of Financial Economics* **1**, pp. 97–103.

Goll, T. and Kallsen, J. (2000). Optimal portfolios for logarithmic utility, *Stochastic Processes and their Application* **89**, 1, pp. 31–48.

Goll, T. and Kallsen, J. (2003). A complete explicit solution to the log-optimal portfolio problem, *Advances in Applied Probability* **13**, 2, pp. 774–779.

Gordon, M. J., Paradis, G. and Rorke, C. (1972). Experimental evidence on alternative portfolio decision rules, *American Economic Review* **62**, 1/2, pp. 107–118.

Grauer, R. R. (1981). A comparison of growth-optimal and mean variance investment policies, *The Journal of Financial and Quantitative Analysis* **16**, 1, pp. 1–21.

Grauer, R. R. and Hakansson, N. H. (1985). Returns on levered, actively managed long-run portfolios of stocks, bonds and bills, 1934-1983, *Financial Analysts Journal* **41**, 5, pp. 24–43.

Gressis, N., Hayya, J. and Philippatos, G. (1974). Multiperiod portfolio efficiency analysis via the geometric mean, *The Financial Review* **9**, 1, pp. 46–63.

Hakansson, N. (1971a). Capital growth and the mean variance approach to portfolio selection, *The Journal of Financial and Quantitative Analysis* **6**, 1, pp. 517–557.

Hakansson, N. (1971b). Multi-period mean-variance analysis: Toward a general theory of portfolio choice, *The Journal of Finance* **26**, 4, pp. 857–884.

Hakansson, N. (1974). Comment on Merton and Samuelson, *Journal of Financial Economics* **1**, 1, p. 95.

Hakansson, N. and Liu, T.-C. (1970). Optimal growth portfolios when yields are serially correlated, *The Review of Economics and Statistics* **52**, 4, pp. 385–394.

Hakansson, N. and Ziemba, W. (1995). Capital growth theory, in R. J. et. al. (ed.), *Handbooks in Operations Research and Management Science*, Vol. 9 (Elsevier Science, Oxford), pp. 65–86.

Harrison, J. M. and Pliska, S. (1981). Martingales and stochastic integrals in the theory of continuous trading, *Stochastic Processes and their Applications* **11**, 3, pp. 215–260.

Hausch, D. B. and Ziemba, W. T. (1990). Arbitrage strategies for cross-track betting on major horse races, *The Journal of Business* **63**, 1, pp. 61–78.

Heath, D., Orey, S., Pestien, V. and Sudderth, W. (1987). Minimizing or maximizing the expected time to reach zero, *SIAM Journal of Control and Optimization* **25**, 1, pp. 195–205.

Heath, D. and Platen, E. (2002a). Consistent pricing and hedging for a modified constant elasticity of variance model, *Quantitative Finance* **2**, 6, pp. 459–467.

Heath, D. and Platen, E. (2002b). Perfect hedging of index derivatives under a minimal market model, *International Journal of Theoretical and Applied Finance* **5**, 7, pp. 757–774.

Heath, D. and Platen, E. (2002c). Pricing and hedging of index derivatives under an alternative asset price model with endogenous stochastic volatility, in *Recent Developments in Mathematical Finance* (World Scientific, Singapore), pp. 117–126.

Heath, D. and Platen, E. (2003). Pricing of index options under a minimal market model with lognormal scaling, *International Journal of Theoretical and Applied Finance* **3**, 6, pp. 442–450.

Henderson, V. and Hobson, D. (2008). Utility indifference pricing - an overview, in R. Carmona (ed.), *Indifference Pricing: Theory and Applications* (Princeton University Press, New Jersey), pp. 44–74.

Hentschel, L. and Long, J. (2004). Numéraire portfolio measures of size and source of gains from international diversification, Working Paper, University of Rochester.

Huang, C.-F. and Litzenberger, R. H. (1988). *Foundations for Finanial Economics* (North Holland Press, Amsterdam).

Hunt, B. (2005a). Feasible high growth investment strategy: Growth-optimal portfolios applied to Dow Jones stocks, *Journal of Asset Management* **6**, 2, pp. 141–157.

Hunt, B. (2005b). Growth-optimal investment strategy efficacy: An application on long run Australian equity data, *Investment Management and Financial Innovations* , 1, pp. 8–22.

Hurd, T. R. (2004). A note on log-optimal portfolios in exponential Levy markets, *Statistics and Decisions* **22**, pp. 225–236.

Ishijima, H. (1999). An empirical analysis on log-utility asset management, *Asia-Pacific Financial Markets* **6**, 3, pp. 253–273.

Ishijima, H., Takano, E. and Taniyama, T. (2004). The growth-optimal asset pricing model under regime switching: With an application to J-REITs, in *Daiwa International Workshop on Financial Engineering*, no.15710120.

Jacod, J. and Shiryaev, A. N. (1987). *Limit Theorems for Stochastic Processes* (Springer Verlag, New York).

Jean, W. H. (1980). The geometric mean and stochastic dominance, *The Journal of Finance* **35**, 1, pp. 151–158.

Karatzas, I. (1989). Optimization problems in the theory of continuous trading, *SIAM J. Control & Optimization* **27**, 6, pp. 1221–1259.

Karatzas, I. and Shreve, S. (1988). *Brownian Motion and Stochastic Calculus* (Springer Verlag, New York).

Kelly, J. L. (1956). A new interpretation of information rate, *Bell System Techn. Journal* **35**, pp. 917–926.

Khanna, A. and Kulldorff, M. (1999). A generalization of the mutual fund theorem, *Finance and Stochastics* **3**, 2, pp. 167–185.

Korn, R., Oertel, F. and Schäl, M. (2003). The numéraire portfolio in financial markets modeled by a multi-dimensional jump diffusion process, *Decisions in Economics and Finance* **26**, 2, pp. 153–166.

Korn, R. and Schäl, M. (1999). On value preserving and growth-optimal portfolios, *Mathematical Methods of Operations Research* **50**, pp. 189–218.

Korn, R. and Schäl, M. (2009). The numéraire portfolio in discrete time: existence, related concepts and applications, *Radon Series for Computational and Applied Mathematics* **8**, pp. 303–326.

Kramkov, D. and Schachermayer, W. (1999). The assymptotic elasticity of utility functions and optimal investment in incomplete markets, *The Annals of Applied Probability* **9**, 3, pp. 904–950.

Kraus, A. and Litzenberger, R. (1975). Market equilibrium in a multiperiod state preference model with logarithmic utility, *The Journal of Finance* **30**, 5, pp. 1213–1227.

Kreps, D. M. (1988). *Notes on the Theory of Choice* (Westview Press/ Boulder, Colorado and London).

Larsen, K. and Zitković, G. (2008). On the semimartingale property via bounded logarithmic utility, *Annals of Finance* **4**, pp. 255–268.

Latané, H. A. (1959). Criteria for choice among risky ventures, *Journal of Political Economy* **67**, pp. 144–155.

Latané, H. A. (1978). The geometric-mean principle revisited - a reply, *Journal of Banking and Finance* **2**, 4, pp. 395–398.

Latané, H. A. and Tuttle, D. L. (1967). Criteria for portfolio building, *Journal of Finance* **22**, 3, pp. 359–373.

Li, Y. (1993). Growth-security investment strategy for long and short runs, *Management Science* **39**, 8, pp. 915–924.

Li, Y., MacLean, L. and Ziemba, W. T. (2005). Time to wealth goals in capital accumulation and the optimal trade-off of growth versus security, *Quantitative Finance* **5**, 4, pp. 343–355.

Loewenstein, M. and Willard, G. A. (2000a). Local martingales, arbitrage and viability. Free Snacks and Cheap Thrills, *Economic Theory* **16**, 1, pp. 135–161.

Loewenstein, M. and Willard, G. A. (2000b). Rational equilibrium asset-pricing bubbles in continuous trading models, *Journal of Economic Theory* **91**, 1, pp. 17–58.

Long, J. B. (1990). The numéraire portfolio, *Journal of Financial Economics* **26**, 1, pp. 29–69.

Luenberger, D. G. (1993). A preference foundation for log mean-variance criteria in portfolio choice problems, *Journal of Economic Dynamics and Control* **17**, 5-6, pp. 887–906.

MacLean, L., Sanegre, R., Zhao, Y. and Ziemba, W. T. (2004). Capital growth with security, *Journal of Economic Dynamics and Control* **28**, 5, pp. 937–954.

Maclean, L. C., Thorp, E. O. and Ziemba, W. T. (2010). *The Kelly Capital Growth Investment Criterion: Theory And Practice* (World Scientific, Singapore).

Maier, S., Peterson, D. and Weide, J. V. (1977a). A Monte Carlo investigation of characteristics of optimal geometric mean portfolios, *The Journal of Financial and Quantitative Analysis* **12**, 2, pp. 215–233.

Maier, S., Peterson, D. and Weide, J. V. (1977b). A strategy which maximizes the geometric mean return on portfolio investments, *Management Science* **23**, 10, pp. 1117–1123.

Markowitz, H. (1952). Portfolio selection, *The Journal of Finance* **7**, 1, pp. 77–91.

Markowitz, H. (1959). *Portfolio Selection. Efficient Diversification of Investments* (John Wiley and Sons, Hoboken).

Markowitz, H. (1972). Investment for the long run, Rodney L. White Center for Financial Research Working Paper no. 20, University of Pensylvania.

Markowitz, H. (1976). Investment for the long run: New evidence for an old rule, *The Journal of Finance* **31**, 5, pp. 1273–1286.

McEnally, R. W. (1986). Latané's bequest: The best of portfolio strategies, *Journal of Portfolio Management* **12**, 2, pp. 21–30.

Mehra, R. and Prescott, E. (1985). The equity premium: A puzzle, *Journal of Monetary Economics* **15**, 2, pp. 145–161.

Menger, K. (1967). The role of uncertainty in economics, in M. Shubik (ed.), *Essays in Mathematical Economics in Honor of Oskar Morgenstern* (Princeton University Press, Princeton), pp. 211–231.

Merton, R. (1969). Lifetime portfolio selection under uncertainty: The continuous time case, *Review of Economics and Statistics* **51**, 3, pp. 247–257.

Merton, R. (1971). Optimum consumption and portfolio rules in a continuous-time model, *Journal of Economic Theory* **3**, 4, pp. 373–413.

Merton, R. (1973). An intertemporal asset pricing model, *Econometrica* **43**, 5, pp. 867–887.

Merton, R. and Samuelson, P. (1974a). Fallacy of the log-normal approximation to optimal portfolio decision-making over many periods, *Journal of Financial Economics* **1**, 1, pp. 67–94.

Merton, R. and Samuelson, P. (1974b). Generalized mean-variance tradeoffs for best perturbation corrections to approximate portfolio decisions, *The Journal of Finance* **29**, 1, pp. 27–40.

Michaud, R. O. (1981). Risk policy and long-term investment, *The Journal of Financial and Quantitative Analysis* **16**, 2, pp. 147–167.

Michaud, R. O. (2003). A practical framework for portfolio choice, *Journal of Investment Management* **1**, 2, pp. 1–16.

Miller, B. L. (1975). Optimal portfolio decision making where the horizon is infinite, *Management Science* **22**, 2, pp. 220–225.

Miller, S. and Platen, E. (2005). A two-factor model for low interest rate regimes, *Asia-Pacific Financial Markets* **11**, 1, pp. 107–133.

Mossin, J. (1968). Optimal multiperiod portfolio choices, *Journal of Business* **41**, 2, pp. 215–229.

Mossin, J. (1972). *Theory of Financial Markets* (Prentice Hall, Englewood Cliffs, NJ).

Nielsen, L. T. and Vassalou, M. (2004). Sharpe ratios and alphas in continuous time, *Journal of Financial and Quantitative Analysis* **39**, 1, pp. 141–152.

Nielsen, L. T. and Vassalou, M. (2006). The instantaneous capital market line, *Economic Theory* **28**, 3, pp. 651–664.

Novikov, A. A. (1973). On moment inequalities and identities for stochastic integrals, *Proc. Second Japan-USSR Symposiom of Probability. Lecture Notes in Mathematics* **330**, pp. 141–152.

Ophir, T. (1978). The geometric-mean principle revisited, *Journal of Banking and Finance* **2**, 1, pp. 103–107.

Ophir, T. (1979). The geometric-mean principle revisited - a reply to a reply, *Journal of Banking and Finance* **3**, 4, pp. 301–303.

Pestien, V. and Sudderth, W. (1985). Continuous-time red and black: How to control a diffusion to a goal, *Mathematics of Operations Research* **10**, 4, pp. 599–611.

Platen, E. (2001). A minimal financial market model, in M. Kohlman and S. Tang (eds.), *Proceedings of Mathematical Finance: Workshop of the Mathematical Finance Research Project*, Trends in Mathematics (Birkhäuser Verlag, Basel), pp. 293–301.

Platen, E. (2002). Arbitrage in continuous complete markets, *Advances in Applied Probability* **34**, 3, pp. 540–558.

Platen, E. (2003). Diversified portfolios in a benchmark framework, Technical Report, University of Technology, Sydney. QFRG Research Paper 87.

Platen, E. (2004a). A benchmark framework for risk management, in J. Akahori, S. Ogawa and S. Watanabe (eds.), *Stoch. Proc. Appl. Math. Finance* (World Scientific, Singapore), pp. 305–335.

Platen, E. (2004b). A class of complete benchmark models with intensity based jumps, *Journal of Applied Probability* **41**, 1, pp. 19–34.

Platen, E. (2004c). Modeling the volatility and expected value of a diversified world index, *International Journal of Theoretical and Applied Finance* **7**, 4, pp. 511–529.

Platen, E. (2004d). Pricing and hedging for incomplete jump diffusion benchmark models, in G. Yin and Q. Zhang (eds.), *Mathematics of Finance: Proceedings of an AMS-IMS-SIAM Joint Summer Research Conference on Mathematics of Finance* (American Mathematical Society, Providence/Snowbird,Utah), pp. 287–301.

Platen, E. (2005a). Diversified portfolios with jumps in a benchmark framework, *Asia-Pacific Financial Markets* **11**, 1, pp. 1–22.

Platen, E. (2005b). Investment for the short and long run, URL `http://www.qfrc.uts.edu.au/research/research_papers/rp163.pdf`, Working Paper, University of Technology Sydney.

Platen, E. (2005c). On the role of the growth-optimal portfolio in finance, *Australian Economic Papers* **44**, 4, pp. 365–388.

Platen, E. (2006a). A benchmark approach to finance, *Mathematical Finance* **16**, 1, pp. 131–151.

Platen, E. (2006b). Capital asset pricing for markets with intensity based jumps, in M. R. Grossinho, A. N. Shiryaev, M. L. Esquivel and P. E. Oliveira (eds.), *International Conference on Stochastic Finance 2004* (Springer, Lisboa, Portugal), pp. 157–182.

Platen, E. and West, J. (2005). A fair pricing approach to weather derivatives, *Asia-Pacific Financial Markets* **11**, 1, pp. 23–53.

Poundstone, W. (2005). *Fortune's Formula: The untold story of the scientific betting system that beat the casinos and Wall street* (Hill and Wang, New York).

Protter, P. (2004). *Stochastic Integration and Differential Equations*, 2nd edn. (Springer Verlag, Berlin).

Pulley, L. (1983). Mean-variance approximations to expected logarithmic utility, *Operations Research* **31**, 4, pp. 685–696.

Revuz, D. and Yor, M. (1991). *Continuous Martingales and Brownian Motion*, 1st edn. (Springer Verlag, Berlin).

Roll, R. (1973). Evidence on the "growth optimum" model, *The Journal of Finance* **28**, 3, pp. 551–566.

Ross, S. A. (1999). Adding risks: Samuelson's fallacy of large numbers revisited, *The Journal of Financial and Quantitative Analysis* **34**, 3, pp. 323–339.

Rotando, L. M. and Thorp, E. (1992). The Kelly criterion and the stock market, *The American Mathematical Monthly* **99**, 10, pp. 922–931.

Rotar, V. (2004). On optimal investment in the long run: Rank dependent expected utility as a bridge between the maximum-expected-log and maximum-expected-utility criteria, Working Paper, San Diego State University.

Roy, A. (1952). Safety first and the holding of assets, *Econometrica* **20**, 3, pp. 431–449.

Rubinstein, M. (1976). The strong case for the generalized logarithmic utility model as the premier model of financial markets, *The Journal of Finance* **31**, 2, pp. 551–571.

Rubinstein, M. (1991). Continuously rebalanced portfolio strategies, *Journal of Portfolio Management* **18**, 1, pp. 78–81.

Samuelson, P. (1963). Risk and uncertainty: A fallacy of large numbers, *Scientia* **57**, 6, pp. 50–55.

Samuelson, P. (1969). Lifetime portfolio selection by dynamical stochastic programming, *The Review of Economics and Statistics* **51**, 3, pp. 239–246.

Samuelson, P. (1971). The "fallacy" of maximizing the geometric mean in long sequences of investing or gambling, *Proceedings of the National Academy of Sciences of the United States of America* **68**, 10, pp. 2493–2496.

Samuelson, P. (1977). St. Petersburg paradoxes: Defanged, dissected and historically described, *Journal of Economic Literature* **15**, 1, pp. 24–55.

Samuelson, P. (1979). Why we should not make mean log of wealth big though years to act are long, *Journal of Banking and Finance* **3**, 4, pp. 305–307.

Samuelson, P. (1984). Additive insurance via the $\sqrt{n}$ law: Domar–Musgrave to the rescue of the Brownian motive. Unpublished manuscript. MIT, Cambridge, MA.

Samuelson, P. (1991). Long-run risk tolerance when equity returns are mean regressing: Pseudoparadoxes and vindication of businessmens risk, in W. C. Brainard, W. D. Nordhaus and H. W. Watts (eds.), *Money, Macroeconomics and Economic Policy* (MIT Press, Cambridge, MA), pp. 181–200.

Schweizer, M. (1995). On the minimal martingale measure and the Föllmer–Schweizer decomposition, *Stochastic Analysis and Applications* **13**, pp. 573–599.

Serva, M. (1999). Optimal lag in dynamic investments, *International Journal of Theoretical and Applied Finance* **2**, 4, pp. 471–481.

Thorp, E. O. (1966). *Beat the Dealer* (Vintage Books, New York).

Thorp, E. O. (1969). Optimal gambling systems for favorable games, *Review of the International Statistical Institute* **37**, 3, pp. 273–293.

Thorp, E. O. (1971). Portfolio choice and the Kelly criterion, *Business and Economics Statistics Section of Proceedings of the American Statistical Association* , pp. 215–224.

Thorp, E. O. (2000). The Kelly criterion in blackjack, sports betting and the stock market, in O. Vancura, J. A. Cornelius and W. R. Eadington (eds.), *Finding the Edge: Mathematical Analysis of Casino Games* (University of Nevada, Reno, NV), pp. 163–213.

Trent, R. H. and Young, W. E. (1969). Geometric mean approximations of individual security and portfolio performance, *The Journal of Financial and Quantitative Analysis* **4**, 2, pp. 179–199.

Treussard, J. (2005). On the validity of risk measures over time: Value-at-risk, conditional tail expectations and the Bodie–Merton–Perold put, URL `http://ideas.repec.org/p/bos/wpaper/wp2005-029.html`, Working paper WP2005-029, Boston University.

Vivian, R. W. (2003). Considering Samuelson's "fallacy of the law of large numbers" without utility, *The South African Journal of Economics* **71**, 2, pp. 363–379.

Williams, J. B. (1936). Speculation and the carryover, *The Quarterly Journal of Economics* **50**, 3, pp. 436–455.

Yan, J., Zhang, Q. and Zhang, S. (2000). Growth-optimal portfolio in a market driven by a jump-diffusion-like process or a levy process, *Annals of Economics and Finance* **1**, 1, pp. 101–116.

Zhao, Y. and Ziemba, W. T. (2003). Intertemporal mean-variance efficiency with a Markovian state price density, URL `http://edoc.hu-berlin.de/series/speps/2003-21/PDF/21.pdf`, Working Paper, University of British Columbia, Vancouver.

Ziemba, W. T. (1972). Note on "optimal growth portfolios when yields are serially correlated", *Journal of Financial and Quantitative Analysis* **7**, 4, pp. 1995–2000.

Ziemba, W. T. (2003). The capital growth - the theory of investment: Part 1, *Wilmott Magazine*, pp. 16–18 URL `www.wilmott.com`.

Ziemba, W. T. (2004). The capital growth - the theory of investment: Part 2, *Wilmott Magazine*, pp. 14–17 URL `www.wilmott.com`.

Ziemba, W. T. (2005). Good and bad properties of the Kelly criterion, *Wilmott Magazine*, pp. 6–9 URL `www.wilmott.com`.

Chapter 2

Empirical Log-Optimal Portfolio Selections: A Survey

László Györfi*, György Ottucsák[†] and András Urbán[‡]

Department of Computer Science and Information Theory,
Budapest University of Technology and Economics,
H-1117, Magyar tudósok körútja 2., Budapest, Hungary.
*gyorfi@cs.bme.hu, [†] oti@cs.bme.hu, [‡] urbi@cs.bme.hu

This chapter provides a survey of discrete-time, multi-period, sequential investment strategies for financial markets. With memoryless assumption about the underlying process generating the asset prices, the best rebalancing is the log-optimal portfolio, which achieves the maximum asymptotic average growth rate. We show some examples (Kelly game, horse racing, St. Petersburg game) illustrating the surprising possibilities for rebalancing. Semi-log-optimal portfolio selection, as a small computational-complexity alternative of the log-optimal portfolio selection, is studied, both theoretically and empirically. For generalized dynamic portfolio selection, when asset prices are generated by a stationary and ergodic process, universally-consistent empirical methods are demonstrated. The empirical performance of the methods is illustrated for NYSE data.

2.1. Introduction

This chapter gives an overview of the investment strategies in financial stock markets inspired by the results of information theory, nonparametric statistics and machine learning. Investment strategies are allowed to use information collected from the market's past and determine, at the beginning of a trading period, a portfolio; that is, a way to distribute their current capital among the available assets. The goal of the investor is to maximize his wealth in the long run without knowing the underlying distribution generating the stock prices. Under this assumption, the asymptotic rate of growth has a well-defined maximum which can be achieved in full knowledge of the underlying distribution generated by the stock prices.

Both static (buy and hold) and dynamic (daily rebalancing) portfolio selections are considered under various assumptions about the behavior of the market process. In case of static portfolio selection, it was shown that every static portfolio asymptotically approximates the growth rate of the best asset in the study. One can achieve larger growth rates with daily rebalancing. With a memoryless assumption about the underlying process generating the asset prices, the log-optimal portfolio achieves the maximum asymptotic average growth rate; that is, the expected value of the logarithm of the return for the best constant-portfolio vector. In order to overcome the high computational-complexity of the log-optimal portfolio, an alternative approximate solutions, semi-log-optimal portfolio selection, is investigated, both theoretically and empirically. Applying recent developments in nonparametric estimation and machine-learning algorithms, for generalized dynamic portfolio selection, when asset prices are generated by a stationary and ergodic process, universally consistent (empirical) methods that achieve the maximum possible growth rate are demonstrated. The spectacular empirical performance of the methods are illustrated for NYSE data.

Consider a market consisting of d assets. The evolution of the market in time is represented by a sequence of price vectors $\mathbf{s}_1, \mathbf{s}_2, \ldots \in \mathbb{R}_+^d$, where

$$\mathbf{s}_n = (s_n^{(1)}, \ldots, s_n^{(d)}),$$

such that the j-th component $s_n^{(j)}$ of $\mathbf{s}_n$ denotes the price of the j-th asset on the n-th trading period. In order to normalize, put $s_0^{(j)} = 1$. $\{\mathbf{s}_n\}$ has an exponential trend:

$$s_n^{(j)} = e^{n W_n^{(j)}} \approx e^{n W^{(j)}},$$

with average growth rate (average yield)

$$W_n^{(j)} \triangleq \frac{1}{n} \ln s_n^{(j)}$$

and with asymptotic average growth rate

$$W^{(j)} \triangleq \lim_{n \to \infty} \frac{1}{n} \ln s_n^{(j)}.$$

The static portfolio selection is a single-period investment strategy. A portfolio vector is denoted by $\mathbf{b} = (b^{(1)}, \ldots b^{(d)})$. (In Chapter 1 of this volume, the components $b^{(j)}$ of this portfolio vector are called fractions and they are denoted by π_j.) The j-th component $b^{(j)}$ of $\mathbf{b}$ denotes the proportion of the investor's capital invested in asset j. We assume that the portfolio vector **b** has nonnegative components and sum to 1, which means that short selling is not permitted. The set of portfolio vectors is denoted by

$$\Delta_d = \left\{ \mathbf{b} = (b^{(1)}, \dots, b^{(d)}); \ b^{(j)} \ge 0, \ \sum_{j=1}^{d} b^{(j)} = 1 \right\}.$$

The aim of static portfolio selection is to achieve $\max_{1 \le j \le d} W^{(j)}$. The static portfolio is an index, for example the S&P 500, such that, at time $n = 0$, we distribute the initial capital S_0 according to a fixed portfolio vector **b**. In other words, if S_n denotes the wealth at the trading period n, then

$$S_n = S_0 \sum_{j=1}^{d} b^{(j)} s_n^{(j)}.$$

Apply the following simple bounds

$$S_0 \max_j b^{(j)} s_n^{(j)} \le S_n \le d S_0 \max_j b^{(j)} s_n^{(j)}.$$

If $b^{(j)} > 0$ for all $j = 1, \dots, d$, then these bounds imply that

$$W \triangleq \lim_{n \to \infty} \frac{1}{n} \ln S_n = \lim_{n \to \infty} \max_j \frac{1}{n} \ln s_n^{(j)} = \max_j W^{(j)}.$$

Thus, any static portfolio selection achieves the growth rate of the best asset in the study, $\max_j W^{(j)}$, and, so, the limit does not depend on the portfolio **b**. In the case of uniform portfolio (uniform index), $b^{(j)} = 1/d$, and the convergence above is from below:

$$S_0 \max_j s_n^{(j)}/d \le S_n \le S_0 \max_j s_n^{(j)}.$$

The rest of this chapter is organized as follows. In Section 2.2, the constantly-rebalanced portfolio is introduced, and the properties of log-optimal portfolio selection are analyzed in the case of a memoryless market. Next, a small computational complexity alternative of the log-optimal portfolio selection, the semi-log optimal portfolio, is introduced. In Section 2.3, the general model of the dynamic portfolio selection is introduced and the basic features of the conditionally log-optimal portfolio selection in the case of a stationary and ergodic market are summarized. Using the principles of nonparametric statistics and machine learning, universally consistent, empirical investment strategies that are able to achieve the maximal asymptotic growth rate are introduced. Experiments on the NYSE data are given in Section 2.3.7.

2.2. Constantly-Rebalanced Portfolio Selection

In order to apply the usual prediction techniques for time series analysis one has to transform the sequence of price vectors $\{\mathbf{s}_n\}$ into a more-or-less stationary sequence of return vectors (price relatives) $\{\mathbf{x}_n\}$ as follows:

$$\mathbf{x}_n = (x_n^{(1)}, \ldots, x_n^{(d)})$$

such that

$$x_n^{(j)} = \frac{s_n^{(j)}}{s_{n-1}^{(j)}} .$$

Thus, the j-th component $x_n^{(j)}$ of the return vector $\mathbf{x}_n$ denotes the amount obtained after investing a unit capital in the j-th asset on the n-th trading period.

With respect to the static portfolio, one can achieve even a higher growth rate for long-run investments, if rebalancing is performed, i.e. if tuning of the portfolio is allowed *dynamically* after each trading period. The dynamic portfolio selection is a multi-period investment strategy, where, at the beginning of each trading period, we can rearrange the wealth among the assets. A representative example of dynamic portfolio selection is the constantly-rebalanced portfolio (CRP), which was introduced and studied by [Kelly (1956)], [Latané (1959)], [Breiman (1961)], [Markowitz (1976)], [Finkelstein and Whitley (1981)], [Móri (1982b)], [Móri and Székely (1982)] and [Barron and Cover (1988)]. For a comprehensive survey, see also Chapter 1 of this volume, Chapters 6 and 15 in [Cover and Thomas (1991)], and Chapter 15 in [Luenberger (1998)].

[Luenberger (1998)] summarizes the main conclusions as follows:

> "Conclusions about multi-period investment situations are not mere variations of single-period conclusions – rather they often *reverse* those earlier conclusions. This makes the subject exciting, both intellectually and in practice. Once the subtleties of multi-period investment are understood, the reward in terms of enhanced investment performance can be substantial."

> "Fortunately the concepts and the methods of analysis for multi-period situation build on those of earlier chapters. Internal rate of return, present value, the comparison principle, portfolio design, and lattice and tree valuation all have natural extensions to general situations. But conclusions such as volatility is 'bad' or diversification is 'good' are no longer universal truths. The story is much more interesting."

In case of CRP, we fix a portfolio vector $\mathbf{b} \in \Delta_d$, i.e., we are concerned with a hypothetical investor who neither consumes nor deposits new cash into his portfolio, but reinvests his portfolio each trading period. In fact, neither short selling nor leverage is allowed. (Concerning short selling and leverage, see Chapter 4 of this volume.) Note that, in this case, the investor has to rebalance his portfolio after each trading day to "corrigate" the daily price shifts of the invested stocks.

Let S_0 denote the investor's initial capital. Then, at the beginning of the first trading period, $S_0 b^{(j)}$ is invested into asset j, and it results in return $S_0 b^{(j)} x_1^{(j)}$. Therefore, at the end of the first trading period, the investor's wealth becomes

$$S_1 = S_0 \sum_{j=1}^{d} b^{(j)} x_1^{(j)} = S_0 \langle \mathbf{b} \,, \mathbf{x}_1 \rangle,$$

where $\langle \cdot \,, \cdot \rangle$ denotes the inner product. For the second trading period, S_1 is the new initial capital

$$S_2 = S_1 \cdot \langle \mathbf{b} \,, \mathbf{x}_2 \rangle = S_0 \cdot \langle \mathbf{b} \,, \mathbf{x}_1 \rangle \cdot \langle \mathbf{b} \,, \mathbf{x}_2 \rangle.$$

By induction, for the trading period n, the initial capital is S_{n-1}. Therefore,

$$S_n = S_{n-1} \langle \mathbf{b} \,, \mathbf{x}_n \rangle = S_0 \prod_{i=1}^{n} \langle \mathbf{b} \,, \mathbf{x}_i \rangle.$$

The asymptotic average growth rate of this portfolio selection is

$$\lim_{n \to \infty} \frac{1}{n} \ln S_n = \lim_{n \to \infty} \left(\frac{1}{n} \ln S_0 + \frac{1}{n} \sum_{i=1}^{n} \ln \langle \mathbf{b} \,, \mathbf{x}_i \rangle \right)$$

$$= \lim_{n \to \infty} \frac{1}{n} \sum_{i=1}^{n} \ln \langle \mathbf{b} \,, \mathbf{x}_i \rangle.$$

Therefore, without loss of generality, one can assume in the sequel that the initial capital $S_0 = 1$.

2.2.1. *Log-Optimal Portfolio for Memoryless Market Process*

If the market process $\{\mathbf{X}_i\}$ is memoryless, i.e., it is a sequence of independent and identically distributed (i.i.d.) random return vectors, then we show that the best constantly-rebalanced portfolio (BCRP) is the log-optimal portfolio

$$\mathbf{b}^* \triangleq \arg\max_{\mathbf{b} \in \Delta_d} \mathbb{E}\{\ln \langle \mathbf{b} \,, \mathbf{X}_1 \rangle\}.$$

This optimality means that, if $S_n^* = S_n(\mathbf{b}^*)$ denotes the capital after day n achieved by a log-optimal portfolio strategy $\mathbf{b}^*$, then, for any portfolio strategy $\mathbf{b}$ with finite $\mathbb{E}\{(\ln \langle \mathbf{b}, \mathbf{X}_1 \rangle)^2\}$ and with capital $S_n = S_n(\mathbf{b})$, and for any memoryless market process $\{\mathbf{X}_n\}_{-\infty}^{\infty}$,

$$\lim_{n\to\infty} \frac{1}{n} \ln S_n \leq \lim_{n\to\infty} \frac{1}{n} \ln S_n^* \quad \text{almost surely}$$

and maximal asymptotic average growth rate is

$$\lim_{n\to\infty} \frac{1}{n} \ln S_n^* = W^* \triangleq \mathbb{E}\{\ln \langle \mathbf{b}^*, \mathbf{X}_1 \rangle\} \quad \text{almost surely.}$$

The proof of optimality is a simple consequence of the strong law of large numbers. Introduce the notation

$$W(\mathbf{b}) = \mathbb{E}\{\ln \langle \mathbf{b}, \mathbf{X}_1 \rangle\}.$$

Then,

$$\begin{aligned}
\frac{1}{n} \ln S_n &= \frac{1}{n} \sum_{i=1}^{n} \ln \langle \mathbf{b}, \mathbf{X}_i \rangle \\
&= \frac{1}{n} \sum_{i=1}^{n} \mathbb{E}\{\ln \langle \mathbf{b}, \mathbf{X}_i \rangle\} + \frac{1}{n} \sum_{i=1}^{n} (\ln \langle \mathbf{b}, \mathbf{X}_i \rangle - \mathbb{E}\{\ln \langle \mathbf{b}, \mathbf{X}_i \rangle\}) \\
&= W(\mathbf{b}) + \frac{1}{n} \sum_{i=1}^{n} (\ln \langle \mathbf{b}, \mathbf{X}_i \rangle - \mathbb{E}\{\ln \langle \mathbf{b}, \mathbf{X}_i \rangle\}).
\end{aligned}$$

The strong law of large numbers implies that

$$\frac{1}{n} \sum_{i=1}^{n} (\ln \langle \mathbf{b}, \mathbf{X}_i \rangle - \mathbb{E}\{\ln \langle \mathbf{b}, \mathbf{X}_i \rangle\}) \to 0 \quad \text{almost surely.}$$

Therefore,

$$\lim_{n\to\infty} \frac{1}{n} \ln S_n = W(\mathbf{b}) = \mathbb{E}\{\ln \langle \mathbf{b}, \mathbf{X}_1 \rangle\} \quad \text{almost surely.}$$

Similarly,

$$\lim_{n\to\infty} \frac{1}{n} \ln S_n^* = W(\mathbf{b}^*) = \max_{\mathbf{b}} W(\mathbf{b}) \quad \text{almost surely.}$$

We have to emphasize the basic conditions of the model; assume that

(1) the assets are arbitrarily divisible, and they are available for buying and for selling in unbounded quantities at the current price in any given trading period,

(2) there are no transaction costs,

(3) the behavior of the market is not affected by the actions of the investor using the strategy under investigation.

Omitting assumption (2), see Chapter 3 of this volume. For memoryless or Markovian market process, optimal strategies have been introduced if the distributions of the market process are known. For the time being, there is no asymptotically optimal, empirical algorithm which accounts for the proportional transaction cost. Condition (3) means that the market is inefficient.

The principle of log-optimality has the important consequence that

$$S_n(\mathbf{b}) \quad \text{is not close to} \quad \mathbb{E}\{S_n(\mathbf{b})\}.$$

However we can prove a bit more. The optimality property proved above means that, for any $\delta > 0$, the event

$$\left\{ -\delta < \frac{1}{n} \ln S_n(\mathbf{b}) - \mathbb{E}\{\ln \langle \mathbf{b}, \mathbf{X}_1 \rangle\} < \delta \right\}$$

has probability close to 1 if n is large enough. On the one hand, we have that

$$\left\{ -\delta < \frac{1}{n} \ln S_n(\mathbf{b}) - \mathbb{E}\{\ln \langle \mathbf{b}, \mathbf{X}_1 \rangle\} < \delta \right\}$$

$$= \left\{ -\delta + \mathbb{E}\{\ln \langle \mathbf{b}, \mathbf{X}_1 \rangle\} < \frac{1}{n} \ln S_n(\mathbf{b}) < \delta + \mathbb{E}\{\ln \langle \mathbf{b}, \mathbf{X}_1 \rangle\} \right\}$$

$$= \left\{ e^{n(-\delta + \mathbb{E}\{\ln\langle \mathbf{b}, \mathbf{X}_1\rangle\})} < S_n(\mathbf{b}) < e^{n(\delta + \mathbb{E}\{\ln\langle \mathbf{b}, \mathbf{X}_1\rangle\})} \right\}.$$

Therefore,

$$S_n(\mathbf{b}) \quad \text{is close to} \quad e^{n\mathbb{E}\{\ln\langle \mathbf{b}, \mathbf{X}_1\rangle\}}.$$

On the other hand,

$$\mathbb{E}\{S_n(\mathbf{b})\} = \mathbb{E}\left\{ \prod_{i=1}^{n} \langle \mathbf{b}, \mathbf{X}_i \rangle \right\} = \prod_{i=1}^{n} \langle \mathbf{b}, \mathbb{E}\{\mathbf{X}_i\} \rangle = e^{n \ln\langle \mathbf{b}, \mathbb{E}\{\mathbf{X}_1\}\rangle}.$$

By the Jensen inequality,

$$\ln \langle \mathbf{b}, \mathbb{E}\{\mathbf{X}_1\} \rangle > \mathbb{E}\{\ln \langle \mathbf{b}, \mathbf{X}_1 \rangle\}.$$

Therefore,

$$S_n(\mathbf{b}) \quad \text{is much less than} \quad \mathbb{E}\{S_n(\mathbf{b})\}.$$

Not knowing this fact, one can apply a naïve approach

$$\arg\max_{\mathbf{b}} \mathbb{E}\{S_n(\mathbf{b})\}.$$

Because of

$$\mathbb{E}\{S_n(\mathbf{b})\} = \langle \mathbf{b}, \mathbb{E}\{\mathbf{X}_1\}\rangle^n,$$

this naïve approach has the equivalent form

$$\arg\max_{\mathbf{b}} \mathbb{E}\{S_n(\mathbf{b})\} = \arg\max_{\mathbf{b}} \langle \mathbf{b}, \mathbb{E}\{\mathbf{X}_1\}\rangle,$$

which is called the mean approach. It is easy to see that $\arg\max_{\mathbf{b}} \langle \mathbf{b}, \mathbb{E}\{\mathbf{X}_1\}\rangle$ is a portfolio vector having 1 at the position where the vector $\mathbb{E}\{\mathbf{X}_1\}$ has the largest component.

In his seminal paper, [Markowitz (1952)] realized that the mean approach is inadequate, i.e., it is a dangerous portfolio. In order to avoid this difficulty, he suggested a diversification, which is called the mean-variance portfolio, such that

$$\widetilde{\mathbf{b}} = \arg\max_{\mathbf{b}:\mathbf{Var}(\langle \mathbf{b}, \mathbf{X}_1\rangle)\leq\lambda} \langle \mathbf{b}, \mathbb{E}\{\mathbf{X}_1\}\rangle,$$

where $\lambda > 0$ is the investor's risk-aversion parameter.

For appropriate choice of λ, the performance (average growth rate) of $\widetilde{\mathbf{b}}$ can be close to the performance of the optimal $\mathbf{b}^*$. However, a good choice of λ depends on the (unknown) distribution of the return vector $\mathbf{X}$.

The calculation of $\widetilde{\mathbf{b}}$ is a quadratically constrained linear programming problem, where a linear function is maximized under quadratic constraints.

In order to calculate the log-optimal portfolio $\mathbf{b}^*$, one has to know the distribution of $\mathbf{X}_1$. If this distribution is unknown, then the empirical log-optimal portfolio can be defined by

$$\mathbf{b}_n^* = \arg\max_{\mathbf{b}} \frac{1}{n} \sum_{i=1}^{n} \ln \langle \mathbf{b}, \mathbf{X}_i\rangle$$

with linear constraints

$$\sum_{j=1}^{d} b^{(j)} = 1 \quad \text{and} \quad 0 \leq b^{(j)} \leq 1 \quad j = 1,\dots,d.$$

The behavior of the empirical portfolio $\mathbf{b}_n^*$ and its modifications was studied by [Móri (1984, 1986)] and [Morvai (1991, 1992)].

The calculation of $\mathbf{b}_n^*$ is a nonlinear programming (NLP) problem. [Cover (1984)] introduced an algorithm for calculating $\mathbf{b}_n^*$. An alternative possibility is the software routine DONLP2 of [Spellucci (1999)]. The routine is based on a sequential quadratic programming method, which computes sequentially a local solution of NLP by solving a quadratic programming problem, and estimates the global maximum according to these local maximums.

2.2.2. *Examples for the Constantly-Rebalanced Portfolio*

Next we show some examples of portfolio games.

Example 2.1. (Kelly Game [Kelly (1956)]) Consider the example of $d = 2$ and $\mathbf{X} = (X^{(1)}, X^{(2)})$, such that the first component $X^{(1)}$ of the return vector $\mathbf{X}$ is the payoff of the Kelly game

$$X^{(1)} = \begin{cases} 2 & \text{with probability } 1/2, \\ 1/2 & \text{with probability } 1/2, \end{cases} \tag{2.1}$$

and the second component $X^{(2)}$ of the return vector $\mathbf{X}$ is the cash:

$$X^{(2)} = 1.$$

Obviously, the cash has zero growth rate. Use the expectation of the first component

$$\mathbb{E}\{X^{(1)}\} = 1/2 \cdot (2 + 1/2) = 5/4 > 1$$

and assume that we are given an i.i.d. sequence of Kelly payoffs $\{X_i^{(1)}\}_{i=1}^{\infty}$. One can introduce the sequential Kelly game $S_n^{(1)}$, such that there is a reinvestment

$$S_n^{(1)} = \prod_{i=1}^{n} X_i^{(1)}.$$

The i.i.d. property of the payoffs $\{X_i^{(1)}\}_{i=1}^{\infty}$ implies that

$$\mathbb{E}\{S_n^{(1)}\} = \mathbb{E}\left\{\prod_{i=1}^{n} X_i^{(1)}\right\} = (5/4)^n. \tag{2.2}$$

Therefore, $\mathbb{E}\{S_n^{(1)}\}$ grows exponentially. However, it does not imply that the random variable $S_n^{(1)}$ grows exponentially as well. Let us calculate the growth rate $W^{(1)}$:

$$W^{(1)} \triangleq \lim_{n \to \infty} \frac{1}{n} \ln S_n^{(1)} = \lim_{n \to \infty} \frac{1}{n} \sum_{i=1}^{n} \ln X_i^{(1)} = \mathbb{E}\{\ln X^{(1)}\}$$
$$= 1/2 \ln 2 + 1/2 \ln(1/2) = 0,$$

almost surely, which means that the first component $X^{(1)}$ of the return vector $\mathbf{X}$ also has zero growth rate.

The following viewpoint may help explain this property, which may, at first sight, seem surprising. First, we write the evolution of the wealth of the sequential Kelly game as follows: let $S_n^{(1)} = 2^{2B(n,1/2)-n}$, where $B(n,1/2)$ is a binomial random variable with parameters $(n,1/2)$ (it is easy to check by choosing $n = 1$ – we then return back to the one-step performance of the game). Now, we write, according to the Moivre–Laplace theorem (a special case of the central-limit theorem for binomial distribution)

$$\mathbb{P}\left(\frac{2B(n,1/2) - n}{\sqrt{\mathbf{Var}(2B(n,1/2))}} \leq x \right) \simeq \phi(x),$$

where $\phi(x)$ is the cumulative distribution function of the standard normal distribution. Rearranging the left-hand side we have

$$\mathbb{P}\left(\frac{2B(n,1/2) - n}{\sqrt{\mathbf{Var}(2B(n,1/2))}} \leq x \right) = \mathbb{P}\left(2B(n,1/2) - n \leq x\sqrt{n} \right)$$
$$= \mathbb{P}\left(2^{2B(n,1/2)-n} \leq 2^{x\sqrt{n}} \right)$$
$$= \mathbb{P}\left(S_n^{(1)} \leq 2^{x\sqrt{n}} \right)$$

that is,

$$\mathbb{P}\left(S_n^{(1)} \leq 2^{x\sqrt{n}} \right) \simeq \phi(x) .$$

Now, choose x_ε such that $\phi(x_\varepsilon) = 1 - \varepsilon$. Then

$$\mathbb{P}\left(S_n^{(1)} \leq 2^{x_\varepsilon \sqrt{n}} \right) \simeq 1 - \varepsilon$$

and, for a fixed $\varepsilon > 0$ let n_0 be such that

$$2^{x_\varepsilon \sqrt{n}} < \mathbb{E}S_n^{(1)} = \left(\frac{5}{4} \right)^n .$$

We then have, for all $n > n_0$,

$$\mathbb{P}\left(S_n^{(1)} \geq \mathbb{E}S_n^{(1)} \right) \leq \mathbb{P}\left(S_n^{(1)} \geq 2^{x_\varepsilon \sqrt{n}} \right) \simeq \varepsilon.$$

It means that most of the values of $S_n^{(1)}$ are far smaller than their expected values $\mathbb{E}S_n^{(1)}$ (see in Figure 2.1).

Now, let us return to the original problem and calculate the log-optimal portfolio for this return vector, where both components have zero growth rate. The portfolio vector has the form

$$\mathbf{b} = (b, 1 - b).$$

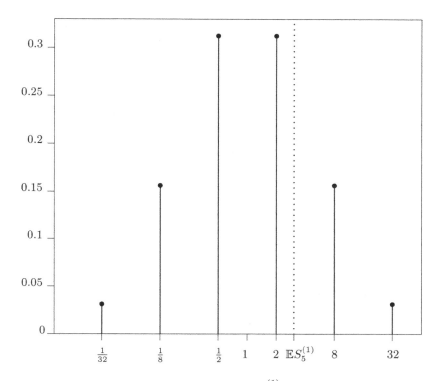

Fig. 2.1.   The distribution of $S_n^{(1)}$ for $n = 5$.

Then,

$$W(\mathbf{b}) = \mathbb{E}\{\ln \langle \mathbf{b}, \mathbf{X}\rangle\}$$
$$= 1/2 \left(\ln(2b + (1 - b)) + \ln(b/2 + (1 - b))\right)$$
$$= 1/2 \ln[(1 + b)(1 - b/2)].$$

One can check that $W(\mathbf{b})$ has a maximum for $b = 1/2$, so the log-optimal portfolio is

$$\mathbf{b}^* = (1/2, 1/2)$$

and the asymptotic average growth rate is

$$W^* = \mathbb{E}\{\ln \langle \mathbf{b}^*, \mathbf{X}\rangle\} = 1/2 \ln(9/8) = 0.059,$$

which is positive.

Example 2.2. Consider the example of $d = 3$ and $\mathbf{X} = (X^{(1)}, X^{(2)}, X^{(3)})$, such that the first and the second components of the return vector $\mathbf{X}$ are Kelly payoffs of form (2.1) and are independent, while the third component is the cash. One can show that the log-optimal portfolio is

$$\mathbf{b}^* = (0.46, 0.46, 0.08)$$

and the maximum asymptotic average growth rate is

$$W^* = \mathbb{E}\{\ln \langle \mathbf{b}^*, \mathbf{X} \rangle\} = 0.112.$$

Example 2.3. Consider the example of $d > 3$ and $\mathbf{X} = (X^{(1)}, X^{(2)}, \ldots, X^{(d)})$, such that the first $d - 1$ independent components of the return vector $\mathbf{X}$ are Kelly payoffs of form (2.1), while the last component is the cash. One can show that the log-optimal portfolio is

$$\mathbf{b}^* = (1/(d-1), \ldots, 1/(d-1), 0),$$

which means that, for $d > 3$, the cash has zero weight according to the log-optimal portfolio. Let N denote the number of components of $\mathbf{X}$ equal to 2. Then, N is binomially distributed with parameters $(d - 1, 1/2)$, and

$$\ln \langle \mathbf{b}^*, \mathbf{X} \rangle = \ln \left(\frac{2N + (d - 1 - N)/2}{d - 1} \right) = \ln \left(\frac{3N}{2(d-1)} + \frac{1}{2} \right).$$

Therefore,

$$W^* = \mathbb{E}\{\ln \langle \mathbf{b}^*, \mathbf{X} \rangle\} = \mathbb{E} \left\{ \ln \left(\frac{3N}{2(d-1)} + \frac{1}{2} \right) \right\}.$$

For $d = 4$, the formula implies that the maximum asymptotic average growth rate is

$$W^* = \mathbb{E}\{\ln \langle \mathbf{b}^*, \mathbf{X} \rangle\} = 0.152,$$

while, for $d \to \infty$,

$$W^* = \mathbb{E}\{\ln \langle \mathbf{b}^*, \mathbf{X} \rangle\} \to \ln(5/4) = 0.223,$$

which means that

$$S_n \approx e^{nW^*} = (5/4)^n,$$

so, with many such Kelly components,

$$S_n \approx \mathbb{E}\{S_n\}$$

(cf. Equation (2.2)).

Example 2.4. (Horse racing [Cover and Thomas (1991)]) Consider the example of horse racing with d horses in a race. Assume that horse j wins with probability p_j. The payoff is denoted by o_j, which means that investing \$1 on horse j results in o_j if it wins, otherwise \$0. Then, the return vector is of the form

$$\mathbf{X} = (0, \ldots, 0, o_j, 0, \ldots, 0)$$

if horse j wins. For repeated races, the return vectors are i.i.d., and so it is a CRP problem. Let us calculate the expected log-return

$$W(\mathbf{b}) = \mathbb{E}\{\ln \langle \mathbf{b}, \mathbf{X} \rangle\} = \sum_{j=1}^{d} p_j \ln(b^{(j)} o_j) = \sum_{j=1}^{d} p_j \ln b^{(j)} + \sum_{j=1}^{d} p_j \ln o_j.$$

Therefore,

$$\arg\max_{\mathbf{b}} \mathbb{E}\{\ln \langle \mathbf{b}, \mathbf{X} \rangle\} = \arg\max_{\mathbf{b}} \sum_{j=1}^{d} p_j \ln b^{(j)}.$$

In order to solve the optimization problem

$$\arg\max_{\mathbf{b}} \sum_{j=1}^{d} p_j \ln b^{(j)},$$

we introduce the Kullback–Leibler divergence of the distributions $\mathbf{p}$ and $\mathbf{b}$:

$$\mathrm{KL}(\mathbf{p}, \mathbf{b}) = \sum_{j=1}^{d} p_j \ln \frac{p_j}{b^{(j)}}.$$

The basic property of the Kullback–Leibler divergence is that

$$\mathrm{KL}(\mathbf{p}, \mathbf{b}) \geq 0,$$

and is equal to zero if and only if the two distributions are equal. The proof of this property is simple:

$$\mathrm{KL}(\mathbf{p}, \mathbf{b}) = -\sum_{j=1}^{d} p_j \ln \frac{b^{(j)}}{p_j} \geq -\sum_{j=1}^{d} p_j \left(\frac{b^{(j)}}{p_j} - 1 \right) = -\sum_{j=1}^{d} b^{(j)} + \sum_{j=1}^{d} p_j = 0.$$

This inequality implies that

$$\arg\max_{\mathbf{b}} \sum_{j=1}^{d} p_j \ln b^{(j)} = \mathbf{p}.$$

Surprisingly, the log-optimal portfolio is independent of the payoffs, and

$$W^* = \sum_{j=1}^{d} p_j \ln(p_j o_j).$$

Knowing the distribution **p**, the usual choice of payoffs is

$$o_j = \frac{1}{p_j},$$

and then

$$W^* = 0.$$

It means that, for this choice of payoffs, any gambling strategy has negative growth rate.

Example 2.5. (Sequential St. Petersburg Games) Consider the simple St. Petersburg game, where the player invests \$1 and a fair coin is tossed until a tail first appears, ending the game. If the first tail appears in step k then the the payoff X is 2^k and the probability of this event is 2^{-k}:

$$\mathbb{P}\{X = 2^k\} = 2^{-k}.$$

Since $\mathbb{E}\{X\} = \infty$, this game has delicate properties; see [Aumann (1977)], [Bernoulli (1954)], [Durand (1957)], [Haigh (1999)], [Martin (2004)], [Menger (1934)], [Rieger and Wang (2006)] and [Samuelson (1960)]. In the literature, usually the repeated St. Petersburg game (also called the iterated St. Petersburg game) means a multi-period game such that it is a sequence of simple St. Petersburg games, where, in n-th round, the player invest \$1 and gets the payoff X_n. Assume that the sequence $\{X_n\}_{n=1}^{\infty}$ is independent and identically distributed. After n rounds, the player's wealth in the repeated game is

$$\tilde{S}_n = \sum_{i=1}^{n} X_i.$$

Then,

$$\lim_{n \to \infty} \frac{\tilde{S}_n}{n \log_2 n} = 1$$

in probability, where $\log_2$ denotes the logarithm to base 2 (cf. [Feller (1945)]). Moreover,

$$\liminf_{n \to \infty} \frac{\tilde{S}_n}{n \log_2 n} = 1 \quad \text{almost surely}$$

and

$$\limsup_{n \to \infty} \frac{\tilde{S}_n}{n \log_2 n} = \infty \quad \text{almost surely}$$

(cf. [Chow and Robbins (1961)]). Introducing the notation for the largest payoff

$$X_n^* = \max_{1 \le i \le n} X_i$$

and for the sum with the largest payoff withheld

$$S_n^* = \tilde{S}_n - X_n^*,$$

one has that

$$\lim_{n \to \infty} \frac{S_n^*}{n \log_2 n} = 1 \quad \text{almost surely}$$

(cf. [Csörgő and Simons (1996)]). According to the previous results, $\tilde{S}_n \approx n \log_2 n$. Similarly to [Bell and Cover (1980)], we introduce a multi-period game, called a sequential St. Petersburg game, having exponential growth. The sequential St. Petersburg game means that the player starts with initial capital $S_0 = \$1$, and there is an independent sequence of simple St. Petersburg games, and, for each simple, game the player reinvests his capital. If $S_{n-1}^{(c)}$ is the capital after the $(n-1)$-th simple game, then the invested capital is $S_{n-1}^{(c)}(1-c)$, while $S_{n-1}^{(c)}c$ is the proportional cost of the simple game with commission factor $0 < c < 1$. It means that, after the n-th round, the capital is

$$S_n^{(c)} = S_{n-1}^{(c)}(1-c)X_n = S_0(1-c)^n \prod_{i=1}^{n} X_i = (1-c)^n \prod_{i=1}^{n} X_i.$$

Because of its multiplicative definition, $S_n^{(c)}$ has an exponential trend:

$$S_n^{(c)} = e^{n W_n^{(c)}} \approx e^{n W^{(c)}},$$

with average growth rate

$$W_n^{(c)} \triangleq \frac{1}{n} \ln S_n^{(c)}$$

and with asymptotic average growth rate

$$W^{(c)} \triangleq \lim_{n \to \infty} \frac{1}{n} \ln S_n^{(c)}.$$

Let us calculate the asymptotic average growth rate. Because of

$$W_n^{(c)} = \frac{1}{n} \ln S_n^{(c)} = \frac{1}{n} \left(n \ln(1-c) + \sum_{i=1}^{n} \ln X_i \right),$$

the strong law of large numbers implies that

$$W^{(c)} = \ln(1-c) + \lim_{n \to \infty} \frac{1}{n} \sum_{i=1}^{n} \ln X_i = \ln(1-c) + \mathbb{E}\{\ln X_1\} \quad \text{almost surely,}$$

so $W^{(c)}$ can be calculated via expected log-utility (cf. [Kenneth (1974)]). A commission factor c is called fair if

$$W^{(c)} = 0,$$

so the growth rate of the sequential game is 0. Let us calculate the fair c:

$$\ln(1-c) = -\mathbb{E}\{\ln X_1\} = -\sum_{k=1}^{\infty} k \ln 2 \cdot 2^{-k} = -2 \ln 2,$$

i.e.,

$$c = 3/4.$$

[Győrfi and Kevei (2009)] studied the portfolio game, where a fraction of the capital is invested in the simple fair St. Petersburg game and the rest is kept in cash. This is the model of the CRP. Fix a portfolio vector $\mathbf{b} = (b, 1-b)$, with $0 \le b \le 1$. Let $S_0 = 1$ denote the player's initial capital. Then, at the beginning of the portfolio game, $S_0 b = b$ is invested into the fair game, and it results in return $bX_1/4$, while $S_0(1-b) = 1-b$ remains in cash. Therefore, after the first round of the portfolio game, the player's wealth becomes

$$S_1 = S_0(bX_1/4 + (1-b)) = b(X_1/4 - 1) + 1.$$

For the second portfolio game, S_1 is the new initial capital

$$S_2 = S_1(b(X_2/4 - 1) + 1) = (b(X_1/4 - 1) + 1)(b(X_2/4 - 1) + 1).$$

By induction, for the n-th portfolio game, the initial capital is S_{n-1}. Therefore,

$$S_n = S_{n-1}(b(X_n/4 - 1) + 1) = \prod_{i=1}^{n} (b(X_i/4 - 1) + 1).$$

The asymptotic average growth rate of this portfolio game is

$$W(b) \triangleq \lim_{n \to \infty} \frac{1}{n} \log_2 S_n$$

$$= \lim_{n \to \infty} \frac{1}{n} \sum_{i=1}^{n} \log_2(b(X_i/4 - 1) + 1)$$

$$\to \mathbb{E}\{\log_2(b(X_1/4 - 1) + 1)\} \quad \text{almost surely.}$$

The function ln is concave, therefore $W(b)$ is also concave, so $W(0) = 0$ (keep everything in cash) and $W(1) = 0$ (the simple St. Petersburg game is fair) imply that, for all $0 < b < 1$ and $W(b) > 0$. Let us calculate

$$\max_b W(b).$$

We have that

$$W(b) = \sum_{k=1}^{\infty} \log_2(b(2^k/4 - 1) + 1) \cdot 2^{-k}$$

$$= \log_2(1 - b/2) \cdot 2^{-1} + \sum_{k=3}^{\infty} \log_2(b(2^{k-2} - 1) + 1) \cdot 2^{-k}.$$

One can show that $\mathbf{b}^* = (0.385, 0.615)$ and $W^* = 0.149$.

Example 2.6. We can extend Example 2.5 such that, in each round, there are d St. Petersburg components, i.e., the return vector has the form

$$\mathbf{X} = (X^{(1)}, \dots, X^{(d)}, X^{(d+1)}) = (X_1/4, \dots, X_d/4, 1)$$

($d \geq 1$), where the first d i.i.d. components of $\mathbf{X}$ are fair St. Petersburg payoffs, while the last component is the cash. For $d = 2$, $\mathbf{b}^* = (0.364, 0.364, 0.272)$. For $d \geq 3$, the best portfolio is the uniform portfolio such that the cash has zero weight:

$$\mathbf{b}^* = (1/d, \dots, 1/d, 0)$$

and the asymptotic average growth rate is

$$W_d^* = \mathbb{E}\left\{ \log_2\left(\frac{1}{4d} \sum_{i=1}^{d} X_i \right) \right\}.$$

Table 2.1 lists the first few values.
[Györfi and Kevei (2011)] proved that

$$W_d^* \approx \log_2 \log_2 d - 2 + \frac{\log_2 \log_2 d}{\ln 2 \log_2 d},$$

which results in some figures for large d (see Table 2.2).

Table 2.1.   Numerical results.

d	1	2	3	4	5	6	7	8
W_d^*	0.149	0.289	0.421	0.526	0.606	0.669	0.721	0.765

Table 2.2.   Simulation results.

d	8	16	32	64
W_d^*	0.76	0.97	1.17	1.35

2.2.3. Semi-Log-Optimal Portfolio

[Roll (1973)], [Pulley (1983)] and [Vajda (2006)] suggested an approximation of $\mathbf{b}^*$ and $\mathbf{b}_n^*$ using

$$h(z) \triangleq z - 1 - \frac{1}{2}(z-1)^2,$$

which is the second-order Taylor expansion of the function $\ln z$ at $z = 1$. Then, the semi-log-optimal portfolio selection is

$$\bar{\mathbf{b}} = \arg\max_{\mathbf{b}} \mathbb{E}\{h(\langle \mathbf{b}, \mathbf{X}_1 \rangle)\},$$

and the empirical semi-log-optimal portfolio is

$$\bar{\mathbf{b}}_n = \arg\max_{\mathbf{b}} \frac{1}{n}\sum_{i=1}^{n} h(\langle \mathbf{b}, \mathbf{x}_i \rangle).$$

In order to compute $\mathbf{b}_n^*$, one has to make an optimization over $\mathbf{b}$. In each optimization step, the computational complexity is proportional to n. For $\bar{\mathbf{b}}_n$, this complexity can be reduced. We have that

$$\frac{1}{n}\sum_{i=1}^{n} h(\langle \mathbf{b}, \mathbf{x}_i \rangle) = \frac{1}{n}\sum_{i=1}^{n}(\langle \mathbf{b}, \mathbf{x}_i \rangle - 1) - \frac{1}{2}\frac{1}{n}\sum_{i=1}^{n}(\langle \mathbf{b}, \mathbf{x}_i \rangle - 1)^2.$$

If $\mathbf{1}$ denotes the all-1 vector, then

$$\frac{1}{n}\sum_{i=1}^{n} h(\langle \mathbf{b}, \mathbf{x}_i \rangle) = \langle \mathbf{b}, \mathbf{m} \rangle - \langle \mathbf{b}, \mathbf{Cb} \rangle,$$

where

$$\mathbf{m} = \frac{1}{n}\sum_{i=1}^{n}(\mathbf{x}_i - 1)$$

and

$$C = \frac{1}{2}\frac{1}{n}\sum_{i=1}^{n}(x_i - 1)(x_i - 1)^T.$$

If we calculate the vector m and the matrix C beforehand, then, in each optimization step, the complexity does not depend on n, so the running time for calculating $\bar{b}_n$ is much smaller than for b_n^*. The other advantage of the semi-log-optimal portfolio is that it can be calculated via quadratic programming, which is possible, for example, by using the routine QUAD-PROG++ of [Gaspero (2006)]. This program uses Goldfarb–Idnani dual method for solving quadratic-programming problems [Goldfarb and Idnani (1983)].

2.3. Time-Varying Portfolio Selection

For a general dynamic portfolio selection, the portfolio vector may depend on the past data. As before, $x_i = (x_i^{(1)}, \ldots x_i^{(d)})$ denotes the return vector on trading period i. Let $b = b_1$ be the portfolio vector for the first trading period. For initial capital S_0, we obtain

$$S_1 = S_0 \cdot \langle b_1, x_1 \rangle.$$

For the second trading period, S_1 is the new initial capital, the portfolio vector is $b_2 = b(x_1)$, and

$$S_2 = S_0 \cdot \langle b_1, x_1 \rangle \cdot \langle b(x_1), x_2 \rangle.$$

For the nth trading period, the portfolio vector is $b_n = b(x_1, \ldots, x_{n-1}) = b(x_1^{n-1})$ and

$$S_n = S_0 \prod_{i=1}^{n} \langle b(x_1^{i-1}), x_i \rangle = S_0 e^{nW_n(B)}$$

with average growth rate

$$W_n(B) = \frac{1}{n}\sum_{i=1}^{n} \ln \langle b(x_1^{i-1}), x_i \rangle.$$

2.3.1. *Log-Optimal Portfolio for Stationary Market Process*

The fundamental limits, determined in [Móri (1982a)], [Algoet and Cover (1988)], and [Algoet (1992, 1994)], reveal that the so-called *conditionally* *log-optimal portfolio* $\mathbf{B}^* = \{\mathbf{b}^*(\cdot)\}$ is the best possible choice. More precisely, on trading period n, let $\mathbf{b}^*(\cdot)$ be such that

$$\mathbb{E}\left\{\ln\left\langle\mathbf{b}^*(\mathbf{X}_1^{n-1}),\mathbf{X}_n\right\rangle\big|\mathbf{X}_1^{n-1}\right\} = \max_{\mathbf{b}(\cdot)}\mathbb{E}\left\{\ln\left\langle\mathbf{b}(\mathbf{X}_1^{n-1}),\mathbf{X}_n\right\rangle\big|\mathbf{X}_1^{n-1}\right\}.$$

If $S_n^* = S_n(\mathbf{B}^*)$ denotes the capital achieved by a log-optimal portfolio strategy $\mathbf{B}^*$, after n trading periods, then, for any other investment strategy $\mathbf{B}$ with capital $S_n = S_n(\mathbf{B})$ and with

$$\sup_n \mathbb{E}\left\{(\ln\left\langle\mathbf{b}_n(\mathbf{X}_1^{n-1}),\mathbf{X}_n\right\rangle)^2\right\} < \infty,$$

and for any stationary and ergodic process $\{\mathbf{X}_n\}_{-\infty}^{\infty}$,

$$\limsup_{n\to\infty}\left(\frac{1}{n}\ln S_n - \frac{1}{n}\ln S_n^*\right) \le 0 \quad \text{almost surely} \tag{2.3}$$

and

$$\lim_{n\to\infty}\frac{1}{n}\ln S_n^* = W^* \quad \text{almost surely,}$$

where

$$W^* \triangleq \mathbb{E}\left\{\max_{\mathbf{b}(\cdot)}\mathbb{E}\left\{\ln\left\langle\mathbf{b}(\mathbf{X}_{-\infty}^{-1}),\mathbf{X}_0\right\rangle\big|\mathbf{X}_{-\infty}^{-1}\right\}\right\}$$

is the maximum possible growth rate of any investment strategy. Note that, for memoryless markets, $W^* = \max_{\mathbf{b}}\mathbb{E}\{\ln\left\langle\mathbf{b},\mathbf{X}_0\right\rangle\}$ which shows that, in this case the log-optimal portfolio is the BCRP.

For the proof of this optimality we use the concept of martingale differences:

Definition 2.1. There are two sequences of random variables $\{Z_n\}$ and $\{X_n\}$ such that

- Z_n is a function of $X_1,\ldots,X_n$,

- $\mathbb{E}\{Z_n \mid X_1,\ldots,X_{n-1}\} = 0$ almost surely.

Then, $\{Z_n\}$ is called a martingale difference sequence with respect to $\{X_n\}$.

For martingale difference sequences, there is a strong law of large numbers: If $\{Z_n\}$ is a martingale difference sequence with respect to $\{X_n\}$, and

$$\sum_{n=1}^{\infty}\frac{\mathbb{E}\{Z_n^2\}}{n^2} < \infty$$

then,

$$\lim_{n \to \infty} \frac{1}{n} \sum_{i=1}^{n} Z_i = 0 \quad \text{almost surely}$$

(cf. [Chow (1965)], see also Theorem 3.3.1 in [Stout (1974)]).

In order to be self-contained, we prove a weak law of large numbers for martingale differences. We show that, if $\{Z_n\}$ is a martingale difference sequence with respect to $\{X_n\}$, then $\{Z_n\}$ are uncorrelated. Put $i < j$, then

$$\mathbb{E}\{Z_i Z_j\} = \mathbb{E}\{\mathbb{E}\{Z_i Z_j \mid X_1, \ldots, X_{j-1}\}\}$$
$$= \mathbb{E}\{Z_i \mathbb{E}\{Z_j \mid X_1, \ldots, X_{j-1}\}\} = \mathbb{E}\{Z_i \cdot 0\} = 0.$$

It implies that

$$\mathbb{E}\left\{ \left(\frac{1}{n} \sum_{i=1}^{n} Z_i \right)^2 \right\} = \frac{1}{n^2} \sum_{i=1}^{n} \sum_{j=1}^{n} \mathbb{E}\{Z_i Z_j\} = \frac{1}{n^2} \sum_{i=1}^{n} \mathbb{E}\{Z_i^2\} \to 0$$

if, for example, $\mathbb{E}\{Z_i^2\}$ is a bounded sequence.

One can construct martingale difference sequence as follows: Let $\{Y_n\}$ be an arbitrary sequence such that Y_n is a function of $X_1, \ldots, X_n$. Put

$$Z_n = Y_n - \mathbb{E}\{Y_n \mid X_1, \ldots, X_{n-1}\}.$$

Then, $\{Z_n\}$ is a martingale difference sequence:

- Z_n is a function of $X_1, \ldots, X_n$,
- $\mathbb{E}\{Z_n | X_1, \ldots, X_{n-1}\}$
  $= \mathbb{E}\{Y_n - \mathbb{E}\{Y_n | X_1, \ldots, X_{n-1}\} | X_1, \ldots, X_{n-1}\} = 0 \quad \text{almost surely.}$

Now we can prove optimality of the log-optimal portfolio. Introduce the decomposition

$$\frac{1}{n} \ln S_n = \frac{1}{n} \sum_{i=1}^{n} \ln \langle \mathbf{b}(\mathbf{X}_1^{i-1}), \mathbf{X}_i \rangle$$

$$= \frac{1}{n} \sum_{i=1}^{n} \mathbb{E}\{\ln \langle \mathbf{b}(\mathbf{X}_1^{i-1}), \mathbf{X}_i \rangle \mid \mathbf{X}_1^{i-1}\}$$

$$+ \frac{1}{n} \sum_{i=1}^{n} \left(\ln \langle \mathbf{b}(\mathbf{X}_1^{i-1}), \mathbf{X}_i \rangle - \mathbb{E}\{\ln \langle \mathbf{b}(\mathbf{X}_1^{i-1}), \mathbf{X}_i \rangle \mid \mathbf{X}_1^{i-1}\} \right).$$

The last average is an average of martingale differences, so it tends to zero, almost surely. Similarly,

$$\frac{1}{n} \ln S_n^* = \frac{1}{n} \sum_{i=1}^n \mathbb{E}\{\ln \langle \mathbf{b}^*(\mathbf{X}_1^{i-1}), \mathbf{X}_i \rangle \mid \mathbf{X}_1^{i-1}\}$$

$$+ \frac{1}{n} \sum_{i=1}^n \left(\ln \langle \mathbf{b}^*(\mathbf{X}_1^{i-1}), \mathbf{X}_i \rangle - \mathbb{E}\{\ln \langle \mathbf{b}^*(\mathbf{X}_1^{i-1}), \mathbf{X}_i \rangle \mid \mathbf{X}_1^{i-1}\} \right).$$

Because of the definition of the log-optimal portfolio, we have that

$$\mathbb{E}\{\ln \langle \mathbf{b}(\mathbf{X}_1^{i-1}), \mathbf{X}_i \rangle \mid \mathbf{X}_1^{i-1}\} \le \mathbb{E}\{\ln \langle \mathbf{b}^*(\mathbf{X}_1^{i-1}), \mathbf{X}_i \rangle \mid \mathbf{X}_1^{i-1}\},$$

and the proof is finished.

2.3.2. Empirical Portfolio Selection

The optimality relations proved above give rise to the following definition:

Definition 2.2. An empirical (data driven) portfolio strategy **B** is called *universally consistent* with respect to a class $\mathcal{C}$ of stationary and ergodic processes $\{\mathbf{X}_n\}_{-\infty}^{\infty}$ if, for each process in the class,

$$\lim_{n \to \infty} \frac{1}{n} \ln S_n(\mathbf{B}) = W^* \quad \text{almost surely.}$$

It is not at all obvious that such a universally-consistent portfolio strategy exists. The surprising fact that there exists a strategy, universal with respect to a class of stationary and ergodic processes, was proved by [Algoet (1992)].

Most of the papers dealing with portfolio selections assume that the distributions of the market process are known. If the distributions are unknown, then one can apply a two-stage splitting scheme.

(1) In the first time period the investor collects data, and estimates the corresponding distributions. In this period, there is no any investment.
(2) In the second time period the investor derives strategies from the distribution estimates and performs the investments.

In the sequel we show that there is no need to make any splitting; one can construct sequential algorithms such that the investor can make trading during the whole time period, i.e., estimation and the portfolio selection are made on the whole time period.

Let us recapitulate the definition of log-optimal portfolio:

$$\mathbb{E}\{\ln \langle \mathbf{b}^*(\mathbf{X}_1^{n-1}), \mathbf{X}_n \rangle \mid \mathbf{X}_1^{n-1}\} = \max_{\mathbf{b}(\cdot)} \mathbb{E}\{\ln \langle \mathbf{b}(\mathbf{X}_1^{n-1}), \mathbf{X}_n \rangle \mid \mathbf{X}_1^{n-1}\} \,.$$

For a fixed integer $k > 0$ large enough, we expect that

$$\mathbb{E}\{\ln \langle \mathbf{b}(\mathbf{X}_1^{n-1}), \mathbf{X}_n \rangle \mid \mathbf{X}_1^{n-1}\} \approx \mathbb{E}\{\ln \langle \mathbf{b}(\mathbf{X}_{n-k}^{n-1}), \mathbf{X}_n \rangle \mid \mathbf{X}_{n-k}^{n-1}\}$$

and

$$\mathbf{b}^*(\mathbf{X}_1^{n-1}) \approx \mathbf{b}_k(\mathbf{X}_{n-k}^{n-1}) = \arg\max_{\mathbf{b}(\cdot)} \mathbb{E}\{\ln \langle \mathbf{b}(\mathbf{X}_{n-k}^{n-1}), \mathbf{X}_n \rangle \mid \mathbf{X}_{n-k}^{n-1}\}.$$

Because of stationarity,

$$\begin{aligned}
\mathbf{b}_k(\mathbf{x}_1^k) &= \arg\max_{\mathbf{b}(\cdot)} \mathbb{E}\{\ln \langle \mathbf{b}(\mathbf{X}_{n-k}^{n-1}), \mathbf{X}_n \rangle \mid \mathbf{X}_{n-k}^{n-1} = \mathbf{x}_1^k\} \\
&= \arg\max_{\mathbf{b}(\cdot)} \mathbb{E}\{\ln \langle \mathbf{b}(\mathbf{x}_1^k), \mathbf{X}_{k+1} \rangle \mid \mathbf{X}_1^k = \mathbf{x}_1^k\} \\
&= \arg\max_{\mathbf{b}} \mathbb{E}\{\ln \langle \mathbf{b}, \mathbf{X}_{k+1} \rangle \mid \mathbf{X}_1^k = \mathbf{x}_1^k\} \,,
\end{aligned}$$

which is the maximization of the regression function

$$m_{\mathbf{b}}(\mathbf{x}_1^k) = \mathbb{E}\{\ln \langle \mathbf{b}, \mathbf{X}_{k+1} \rangle \mid \mathbf{X}_1^k = \mathbf{x}_1^k\}.$$

Thus, a possible way for asymptotically-optimal empirical portfolio selection is, based on past data, sequentially estimate the regression function $m_{\mathbf{b}}(\mathbf{x}_1^k)$ and choose the portfolio vector, which maximizes the regression function estimate.

2.3.3. *Regression Function Estimation*

We briefly summarize the basics of nonparametric regression function estimation. Concerning the details, we refer to the book of [Györfi *et al.* (2002)] and to Chapter 5 of this volume. Let Y be a real-valued random variable, and let X denote an observation vector taking values in $\mathbb{R}^d$. The regression function is the conditional expectation of Y given X

$$m(x) = \mathbb{E}\{Y \mid X = x\}.$$

If the distribution of (X, Y) is unknown, then one has to estimate the regression function from data. The data is a sequence of i.i.d. copies of (X, Y)

$$D_n = \{(X_1, Y_1), \ldots, (X_n, Y_n)\}.$$

The regression function estimate is of form

$$m_n(x) = m_n(x, D_n).$$

An important class of estimates is the local averaging estimates

$$m_n(x) = \sum_{i=1}^{n} W_{n,i}(x; X_1, \ldots, X_n) Y_i,$$

where usually the weights $W_{n,i}(x; X_1, \ldots, X_n)$ are nonnegative and sum up to 1. Moreover, $W_{n,i}(x; X_1, \ldots, X_n)$ is relatively large if x is close to X_i, otherwise it is zero.

An example of such an estimate is the *partitioning estimate*. Here, one chooses a finite or countably infinite partition $\mathcal{P}_n = \{A_{n,1}, A_{n,2}, \ldots\}$ of $\mathbb{R}^d$ consisting of cells $A_{n,j} \subseteq \mathbb{R}^d$ and defines, for $x \in A_{n,j}$, the estimate by averaging Y_i's with the corresponding X_i's in $A_{n,j}$, i.e.,

$$m_n(x) = \frac{\sum_{i=1}^{n} I_{\{X_i \in A_{n,j}\}} Y_i}{\sum_{i=1}^{n} I_{\{X_i \in A_{n,j}\}}} \quad \text{for } x \in A_{n,j}, \tag{2.4}$$

where I_A denotes the indicator function of set A. Here, and in the following, we use the convention $\frac{0}{0} = 0$. In order to have consistency, on one hand we need that the cells $A_{n,j}$ should be "small", and, on other hand, the number of non-zero terms in the denominator of (2.4) should be "large". These requirements can be satisfied if the sequences of partition $\mathcal{P}_n$ are asymptotically fine, i.e., if

$$\operatorname{diam}(A) = \sup_{x,y \in A} \|x - y\|$$

denotes the diameter of a set such that $\|\cdot\|$ is the Euclidean norm, then, for each sphere S centered at the origin,

$$\lim_{n \to \infty} \max_{j: A_{n,j} \cap S \neq \emptyset} \operatorname{diam}(A_{n,j}) = 0$$

and

$$\lim_{n \to \infty} \frac{|\{j \,:\, A_{n,j} \cap S \neq \emptyset\}|}{n} = 0.$$

For the partition $\mathcal{P}_n$, the most important example is when the cells $A_{n,j}$ are cubes of volume h_n^d. For a cubic partition, the consistency conditions above mean that

$$\lim_{n \to \infty} h_n = 0 \quad \text{and} \quad \lim_{n \to \infty} n h_n^d = \infty. \tag{2.5}$$

The second example of a local averaging estimate is the *Nadaraya–Watson kernel estimate*. Let $K : \mathbb{R}^d \to \mathbb{R}_+$ be a function called the kernel function, and let $h > 0$ be a bandwidth. The kernel estimate is defined by

$$m_n(x) = \frac{\sum_{i=1}^n K\left(\frac{x-X_i}{h}\right) Y_i}{\sum_{i=1}^n K\left(\frac{x-X_i}{h}\right)} .$$

The kernel estimate is a weighted average of the Y_i, where the weight of Y_i (i.e., the influence of Y_i on the value of the estimate at x) depends on the distance between X_i and x. For the bandwidth $h = h_n$, the consistency conditions are (2.5). If one uses the so-called naïve kernel (or window kernel), $K(x) = I_{\{\|x\| \leq 1\}}$, where $I_{\{\cdot\}}$ denotes the indicator function of the events in the brackets that is, it equals 1 if the event is true and 0 otherwise. Then

$$m_n(x) = \frac{\sum_{i=1}^n I_{\{\|x-X_i\| \leq h\}} Y_i}{\sum_{i=1}^n I_{\{\|x-X_i\| \leq h\}}},$$

i.e., one estimates $m(x)$ by averaging Y_i's such that the distance between X_i and x is not greater than h.

Our final example of local averaging estimates is the *k-nearest-neighbor* (*k-NN*) *estimate*. Here, one determines the k nearest X_i's to x in terms of distance $\|x - X_i\|$ and estimates $m(x)$ by the average of the corresponding Y_i's. More precisely, for $x \in \mathbb{R}^d$, let

$$(X_{(1)}(x), Y_{(1)}(x)), \dots, (X_{(n)}(x), Y_{(n)}(x))$$

be a permutation of

$$(X_1, Y_1), \dots, (X_n, Y_n)$$

such that

$$\|x - X_{(1)}(x)\| \leq \cdots \leq \|x - X_{(n)}(x)\|.$$

The k-NN estimate is defined by

$$m_n(x) = \frac{1}{k} \sum_{i=1}^k Y_{(i)}(x).$$

If $k = k_n \to \infty$ such that $k_n/n \to 0$ then the k-nearest-neighbor regression estimate is consistent.

We use the following correspondence between the general regression estimation and portfolio selection:

$$X \sim \mathbf{X}_1^k,$$

$$Y \sim \ln \langle \mathbf{b}, \mathbf{X}_{k+1} \rangle,$$

$$m(x) = \mathbb{E}\{Y \mid X = x\} \sim m_{\mathbf{b}}(\mathbf{x}_1^k) = \mathbb{E}\{\ln \langle \mathbf{b}, \mathbf{X}_{k+1} \rangle \mid \mathbf{X}_1^k = \mathbf{x}_1^k\}.$$

2.3.4. *Histogram-Based Strategy*

Next we describe the *histogram based strategy* due to [Győrfi and Schäfer (2003)] and denote it by $\mathbf{B}^H$. We first define an infinite array of elementary strategies (the so-called *experts*) $\mathbf{B}^{(k,\ell)} = \{\mathbf{b}^{(k,\ell)}(\cdot)\}$, indexed by the positive integers $k, \ell = 1, 2, \ldots$. Each expert $\mathbf{B}^{(k,\ell)}$ is determined by a period length k and by a partition $\mathcal{P}_\ell = \{A_{\ell,j}\}$, $j = 1, 2, \ldots, m_\ell$ of $\mathbb{R}_+^d$ into m_ℓ disjoint cells. Roughly speaking, an expert discretizes the past's sequence according to the partition $\mathcal{P}_\ell$, and browses through all past appearances of the last-seen discretized k-length string. Then, it designs a fixed portfolio vector optimizing the return for the trading periods following each occurrence of this string.

More precisely, to determine the portfolio on the nth trading period, expert $\mathbf{B}^{(k,\ell)}$ looks at the return vectors $\mathbf{x}_{n-k}, \ldots, \mathbf{x}_{n-1}$ of the last k periods, discretizes this kd-dimensional vector by means of the partition $\mathcal{P}_\ell$, and determines the portfolio vector that is optimal for those past trading periods whose preceding k trading periods have identical discretized return vectors to the present one.

Formally, let G_ℓ be the discretization function corresponding to the partition $\mathcal{P}_\ell$, that is,

$$G_\ell(\mathbf{x}) = j, \text{ if } \mathbf{x} \in A_{\ell,j} .$$

With some abuse of notation, for any n and $\mathbf{x}_1^n \in \mathbb{R}^{dn}$, we write $G_\ell(\mathbf{x}_1^n)$ for the sequence $G_\ell(\mathbf{x}_1), \ldots, G_\ell(\mathbf{x}_n)$. Then, define the expert $\mathbf{B}^{(k,\ell)} = \{\mathbf{b}^{(k,\ell)}(\cdot)\}$ by writing, for each $n > k + 1$,

$$\mathbf{b}^{(k,\ell)}(\mathbf{x}_1^{n-1}) = \arg\max_{\mathbf{b} \in \Delta_d} \prod_{i \in J_{k,l,n}} \langle \mathbf{b}, \mathbf{x}_i \rangle , \tag{2.6}$$

where

$$J_{k,l,n} = \left\{ k < i < n : G_\ell(\mathbf{x}_{i-k}^{i-1}) = G_\ell(\mathbf{x}_{n-k}^{n-1}) \right\} ,$$

if $J_{k,l,n} \neq \emptyset$, where $\emptyset$ denotes the empty set, and uniform $\mathbf{b}_0 = (1/d, \ldots, 1/d)$ otherwise.

The problem left is how to choose k and ℓ. There are two extreme cases:

- small k or small ℓ implies that the corresponding regression estimate has large bias,
- large k and large ℓ implies that usually there are few matching, which results in large variance.

A good, data-driven choice of k and ℓ is possible by using recent techniques from machine learning. In an online sequential machine learning setup, k and ℓ are considered as parameters of the estimates, called experts. The basic idea of online sequential machine learning is the combination of the experts. The combination is an aggregated estimate, where an expert has large weight if its past performance is good (cf. [Cesa-Bianchi and Lugosi (2006)]).

The most appealing combination-type of the experts is exponential weighting due to its beneficial theoretical and practical properties. Combine the elementary portfolio strategies $\mathbf{B}^{(k,\ell)} = \{\mathbf{b}_n^{(k,\ell)}\}$ as follows: let $\{q_{k,\ell}\}$ be a probability distribution on the set of all pairs (k, ℓ), such that for all $k, \ell, q_{k,\ell} > 0$.

For a learning parameter $\eta > 0$, introduce the exponential weights

$$w_{n,k,\ell} = q_{k,\ell} e^{\eta \ln S_{n-1}(\mathbf{B}^{(k,\ell)})}.$$

For $\eta = 1$, it means that

$$w_{n,k,\ell} = q_{k,\ell} e^{\ln S_{n-1}(\mathbf{B}^{(k,\ell)})} = q_{k,\ell} S_{n-1}(\mathbf{B}^{(k,\ell)})$$

and

$$v_{n,k,\ell} = \frac{w_{n,k,\ell}}{\sum_{i,j} w_{n,i,j}}.$$

The combined portfolio $\mathbf{b}$ is defined by

$$\mathbf{b}_n(\mathbf{x}_1^{n-1}) = \sum_{k=1}^{\infty} \sum_{\ell=1}^{\infty} v_{n,k,\ell} \mathbf{b}_n^{(k,\ell)}(\mathbf{x}_1^{n-1}).$$

This combination has a simple interpretation:

$$
\begin{aligned}
S_n(\mathbf{B}^H) &= \prod_{i=1}^{n} \left\langle \mathbf{b}_i(\mathbf{x}_1^{i-1}), \mathbf{x}_i \right\rangle \\
&= \prod_{i=1}^{n} \frac{\sum_{k,\ell} w_{i,k,\ell} \left\langle \mathbf{b}_i^{(k,\ell)}(\mathbf{x}_1^{i-1}), \mathbf{x}_i \right\rangle}{\sum_{k,\ell} w_{i,k,\ell}} \\
&= \prod_{i=1}^{n} \frac{\sum_{k,\ell} q_{k,\ell} S_{i-1}(\mathbf{B}^{(k,\ell)}) \left\langle \mathbf{b}_i^{(k,\ell)}(\mathbf{x}_1^{i-1}), \mathbf{x}_i \right\rangle}{\sum_{k,\ell} q_{k,\ell} S_{i-1}(\mathbf{B}^{(k,\ell)})} \\
&= \prod_{i=1}^{n} \frac{\sum_{k,\ell} q_{k,\ell} S_i(\mathbf{B}^{(k,\ell)})}{\sum_{k,\ell} q_{k,\ell} S_{i-1}(\mathbf{B}^{(k,\ell)})} \\
&= \frac{\sum_{k,\ell} q_{k,\ell} S_n(\mathbf{B}^{(k,\ell)})}{\sum_{k,\ell} q_{k,\ell} S_0(\mathbf{B}^{(k,\ell)})} \\
&= \sum_{k,\ell} q_{k,\ell} S_n(\mathbf{B}^{(k,\ell)}).
\end{aligned}
$$

The strategy $\mathbf{B}^H$ then arises from weighting the elementary portfolio strategies $\mathbf{B}^{(k,\ell)} = \{\mathbf{b}_n^{(k,\ell)}\}$ such that the investor's capital becomes

$$
S_n(\mathbf{B}^H) = \sum_{k,\ell} q_{k,\ell} S_n(\mathbf{B}^{(k,\ell)}). \tag{2.7}
$$

It is shown in [Győrfi and Schäfer (2003)] that the strategy $\mathbf{B}^H$ is universally consistent with respect to the class of all ergodic processes such that $\mathbb{E}\{|\log X^{(j)}|\} < \infty$, for all $j = 1, 2, \ldots, d$, under the following two conditions on the partitions used in the discretization:

(a) the sequence of partitions is nested, that is, any cell of $\mathcal{P}_{\ell+1}$ is a subset of a cell of $\mathcal{P}_\ell$, $\ell = 1, 2, \ldots$,
(b) if $\operatorname{diam}(A) = \sup_{\mathbf{x},\mathbf{y}\in A} \|\mathbf{x} - \mathbf{y}\|$ denotes the diameter of a set, then for any sphere $S \subset \mathbb{R}^d$ centered at the origin,

$$
\lim_{\ell \to \infty} \max_{j:A_{\ell,j}\cap S\neq\emptyset} \operatorname{diam}(A_{\ell,j}) = 0 .
$$

2.3.5. *Kernel-Based Strategy*

[Győrfi et al. (2006)] introduced *kernel-based portfolio selection* strategies. Define an infinite array of experts $\mathbf{B}^{(k,\ell)} = \{\mathbf{b}^{(k,\ell)}(\cdot)\}$, where k, ℓ are positive integers. For fixed positive integers k, ℓ, choose the radius $r_{k,\ell} > 0$

such that, for any fixed k,

$$\lim_{\ell \to \infty} r_{k,\ell} = 0.$$

Then, for $n > k + 1$, define the expert $\mathbf{b}^{(k,\ell)}$ by

$$\mathbf{b}^{(k,\ell)}(\mathbf{x}_1^{n-1}) = \arg\max_{\mathbf{b} \in \Delta_d} \prod_{\left\{ k < i < n : \|\mathbf{x}_{i-k}^{i-1} - \mathbf{x}_{n-k}^{n-1}\| \le r_{k,\ell} \right\}} \langle \mathbf{b}, \mathbf{x}_i \rangle,$$

if the sum is non-void, and $\mathbf{b}_0 = (1/d, \ldots, 1/d)$ otherwise. These experts are mixed as in (2.7).

[Györfi *et al.* (2006)] proved that the portfolio scheme $\mathbf{B}^K = \mathbf{B}$ is universally consistent with respect to the class of all ergodic processes such that $\mathbb{E}\{|\ln X^{(j)}|\} < \infty$, for $j = 1, 2, \ldots d$.

Sketch of the proof: Because of the fundamental limit (2.3), we have to prove that

$$\liminf_{n \to \infty} W_n(\mathbf{B}) = \liminf_{n \to \infty} \frac{1}{n} \ln S_n(\mathbf{B}) \ge W^* \quad \text{almost surely.}$$

We have that

$$W_n(\mathbf{B}) = \frac{1}{n} \ln S_n(\mathbf{B})$$

$$= \frac{1}{n} \ln \left(\sum_{k,\ell} q_{k,\ell} S_n(\mathbf{B}^{(k,\ell)}) \right)$$

$$\ge \frac{1}{n} \ln \left(\sup_{k,\ell} q_{k,\ell} S_n(\mathbf{B}^{(k,\ell)}) \right)$$

$$= \frac{1}{n} \sup_{k,\ell} \left(\ln q_{k,\ell} + \ln S_n(\mathbf{B}^{(k,\ell)}) \right)$$

$$= \sup_{k,\ell} \left(W_n(\mathbf{B}^{(k,\ell)}) + \frac{\ln q_{k,\ell}}{n} \right).$$

Thus,

$$\liminf_{n \to \infty} W_n(\mathbf{B}) \ge \liminf_{n \to \infty} \sup_{k,\ell} \left(W_n(\mathbf{B}^{(k,\ell)}) + \frac{\ln q_{k,\ell}}{n} \right)$$

$$\ge \sup_{k,\ell} \liminf_{n \to \infty} \left(W_n(\mathbf{B}^{(k,\ell)}) + \frac{\ln q_{k,\ell}}{n} \right)$$

$$= \sup_{k,\ell} \liminf_{n \to \infty} W_n(\mathbf{B}^{(k,\ell)})$$

$$= \sup_{k,\ell} \epsilon_{k,\ell}.$$

Because of $\lim_{\ell \to \infty} r_{k,\ell} = 0$, we can show that

$$\sup_{k,\ell} \epsilon_{k,\ell} = \lim_{k \to \infty} \lim_{l \to \infty} \epsilon_{k,\ell} = W^*.$$

2.3.6. Nearest-Neighbor-Based Strategy

Define an infinite array of experts $\mathbf{B}^{(k,\ell)} = \{\mathbf{b}^{(k,\ell)}(\cdot)\}$, where $0 < k, \ell$ are integers. As before, k is the window length of the near past, and, for each ℓ, choose $p_\ell \in (0,1)$, such that

$$\lim_{\ell \to \infty} p_\ell = 0. \tag{2.8}$$

Put

$$\hat{\ell} = \lfloor p_\ell n \rfloor.$$

At a given time instant n, the expert searches for the $\hat{\ell}$ nearest-neighbor (NN) matches in the past. For fixed positive integers k, ℓ $(n > k + \hat{\ell} + 1)$, introduce the set of the $\hat{\ell}$ nearest-neighbor matches:

$$\hat{J}_n^{(k,\ell)} = \{i;\ k+1 \le i \le n \text{ such that } \mathbf{x}_{i-k}^{i-1} \text{ is among the } \hat{\ell} \text{ NNs of } \mathbf{x}_{n-k}^{n-1}$$
$$\text{in } \mathbf{x}_1^k, \ldots, \mathbf{x}_{n-k-1}^{n-2}\}.$$

Define the expert by

$$\mathbf{b}^{(k,\ell)}(\mathbf{x}_1^{n-1}) = \arg\max_{\mathbf{b} \in \Delta_d} \prod_{i \in \hat{J}_n^{(k,\ell)}} \langle \mathbf{b}, \mathbf{x}_i \rangle.$$

That is, $\mathbf{b}_n^{(k,\ell)}$ is a fixed portfolio vector according to the returns following these nearest neighbors. These experts are mixed in the same way as in (2.7).

We say that a tie occurs with probability zero if, for any vector $\mathbf{s} = \mathbf{s}_1^k$, the random variable

$$\|\mathbf{X}_1^k - \mathbf{s}\|$$

has a continuous distribution function.

[Győrfi et al. (2008)] proved the following theorem: Assume (2.8) and that a tie occurs with probability zero. Then, the portfolio scheme $\mathbf{B}^{NN}$ is universally consistent with respect to the class of all stationary and ergodic processes such that $\mathbb{E}\{|\log X^{(j)}|\} < \infty$, for $j = 1, 2, \ldots d$.

2.3.7. *Numerical Results on Empirical Portfolio Selection*

The theoretical results above hold under the condition of stationarity. Obviously, the real data of returns (relative prices) are not stationary, therefore we performed some experiments for New York Stock Exchange (NYSE) data. This section gives numerical results on empirical portfolio selection. At the web page [Gelencsér and Ottucsák (2006)] there are two benchmark data sets from NYSE:

- The first data set consists of daily data of 36 stocks with length 22 years (5651 trading days ending in 1985). More precisely, the data set contains the daily price relatives, that were calculated from the nominal values of the *closing prices* corrected by the dividends and the splits for the whole trading day. This data set has been used for testing portfolio selection in [Cover (1991)], [Singer (1997)], [Györfi *et al.* (2006)], [Györfi *et al.* (2008)] and [Györfi *et al.* (2007)].
- The second data set contains 19 stocks and has length 44 years (11178 trading days ending in 2006) and it was generated in the same way as the previous data set (it was augmented by the last 22 years).

Our experiment is on the *second* data set. To make the analysis feasible, some simplifying assumptions are used that need to be taken into account. We remind the reader we assume that

- the assets are arbitrarily divisible,
- the assets are available for buying or for selling in unbounded quantities at the current price in any given trading period,
- there are no transaction costs (in Chapter 3 of this volume we offer solutions to overcome this problem),
- the behavior of the market is not affected by the actions of the investor using the strategy under investigation.

For the 19 assets in the second data set, the average annual yield (AAY) of the static uniform portfolio (uniform index) is 14%, while the best asset was MORRIS (Philip Morris International Inc.) with AAY 20%. These yields match with theoretical consideration derived in Chapter 1 of this volume (Table 1.1).

Table 2.3 summarizes the numerical results for these 19 assets and for BCRP. The first column of Table 2.3 lists the stock's name and the second column shows the AAY. The third and the fourth columns present the weights of the stocks (the components of the best constant-portfolio vector) using the log-optimal and semi-log-optimal algorithms. Surprisingly, the two portfolio vectors are almost the same; according to the last row, the growth rates are both 20%. The period of 44 years is "too short" to be sure that the growth rate of BCRP is much larger than the growth rate of the best asset. Here one can make the same observation as at the end of Chapter 1 of this volume, i.e., the BCRP is very undiversified; only three assets have a positive weight.

Table 2.3. Comparison of the two algorithms for CRPs.

Stock's name	AAY	BCRP	
		log	Semi-log
AHP	13%	0	0
ALCOA	9%	0	0
AMERB	14%	0	0
COKE	14%	0	0
DOW	12%	0	0
DUPONT	9%	0	0
FORD	9%	0	0
GE	13%	0	0
GM	7%	0	0
HP	15%	0.177	0.178
IBM	10%	0	0
INGER	11%	0	0
JNJ	16%	0	0
KIMBC	13%	0	0
MERCK	15%	0	0
MMM	11%	0	0
MORRIS	20%	0.747	0.746
PANDG	13%	0	0
SCHLUM	15%	0.076	0.076
AAY		20%	20%

For the calculation of the optimal portfolio we use a recursive greedy *gradient algorithm*. Introduce the projection P of a vector $\mathbf{b} = (b^{(1)}, \ldots b^{(d)})$ to Δ_d:

$$P(\mathbf{b}) = \frac{\mathbf{b}}{\sum_{j=1}^{d} b^{(j)}}.$$

Put

$$W_n(\mathbf{b}) = \frac{1}{n} \sum_{i=1}^{n} \log \langle \mathbf{b}, \mathbf{x}_i \rangle \,,$$

and let $\mathbf{e}_j$ be the j-th unit vector, i.e., its j-th component is 1, the other components are 0.

GRADIENT ALGORITHM

Parameters: Number of assets d, initial portfolio $\mathbf{b}_0 = (1/d, \ldots, 1/d)$, $V_0 = W_n(\mathbf{b}_0)$ and step size $\delta > 0$.

At iteration steps $k = 1, 2, 3, \ldots,$

(1) calculate

$$W_n(P(\mathbf{b}_{k-1} + \delta \cdot \mathbf{e}_j)) \quad j = 1, \ldots, d \;;$$

(2) if

$$V_{k-1} \geq \max_j W_n(P(\mathbf{b}_{k-1} + \delta \cdot \mathbf{e}_j))$$

then stop, and the result of the algorithm is $\mathbf{b}_{k-1}$. Otherwise, put

$$V_k = \max_j W_n(P(\mathbf{b}_{k-1} + \delta \cdot \mathbf{e}_j))$$

and

$$\mathbf{b}_k = P(\mathbf{b}_{k-1} + \delta \cdot \mathbf{e}_{j^*}),$$

where

$$j^* = \arg\max_j W_n(P(\mathbf{b}_{k-1} + \delta \cdot \mathbf{e}_j)).$$

Go to (1).

Next, we show experiments for time-varying portfolio selection. One can combine the kernel-based portfolio selection and the principle of the semi-log-optimal algorithm in Section 2.2.3 to give the kernel-based semi-log-optimal portfolio (cf. [Györfi *et al.* (2007)]). We present some numerical results obtained by applying the kernel-based semi-log-optimal algorithm to the second NYSE data set.

The proposed universally-consistent empirical portfolio selection algorithms use an infinite array of experts. In practice, we take a finite array of size $K \times L$. In our experiment, we selected $K = 5$ and $L = 10$. Choose the uniform distribution $\{q_{k,\ell}\} = 1/(KL)$ over the experts in use, and the radius

$$r_{k,l}^2 = 0.0002 \cdot d \cdot k(1 + \ell/10),$$

where $k = 1, \ldots, K$ and $\ell = 1, \ldots, L$.

Table 2.4 summarizes the average annual yield achieved by each expert at the last period when investing one unit for the kernel-based semi-log-optimal portfolio. Experts are indexed by $k = 1 \ldots 5$ in columns and $\ell = 1 \ldots 10$ in rows. The average annual yield of the kernel-based semi-log-optimal portfolio is 31%. According to Table 2.3, MORRIS had the best average annual yield, 20%, while the BCRP had average annual yield 20%, so with the kernel based semi-log-optimal portfolio, we have a spectacular improvement.

Table 2.4. The average annual yields of the individual experts for the kernel strategy.

k ℓ	1	2	3	4	5
1	31%	30%	24%	21%	26%
2	34%	31%	27%	25%	22%
3	35%	29%	26%	24%	23%
4	35%	30%	30%	32%	27%
5	34%	29%	33%	24%	24%
6	35%	29%	28%	24%	27%
7	33%	29%	32%	23%	23%
8	34%	33%	30%	21%	24%
9	37%	33%	28%	19%	21%
10	34%	29%	26%	20%	24%

Another interesting feature of Table 2.4 is that, for any fixed ℓ, the best k is equal to 1, so, as far as empirical portfolio is concerned, the Markovian modelling is appropriate. If the time horizon in the experiment was infinity, the numbers in each fixed row would be monotonically increasing. Here we observe almost the opposite, the explanation being that, in the k-th position of a row the dimension of the optimization problem is $19 \cdot k$, so, for larger k, the dimension is too large with respect to the length of the data, i.e., for larger k there are not enough data to "learn" the best portfolio. Again, the time-varying portfolio is very undiversified, such that the subset of assets with non-zero weight is changing from time to time, which makes the problem of transaction cost challenging.

We performed some experiments using the nearest-neighbor strategy. Again, we take a finite array of size $K \times L$, such that $K = 5$ and $L = 10$. Choose the uniform distribution $\{q_{k,\ell}\} = 1/(KL)$ over the experts in use. Table 2.5 summarizes the average annual yield achieved by each expert at the last period when investing one unit for the nearest-neighbor portfolio strategy. Experts are indexed by $k = 1 \ldots 5$ in columns and $\ell = 50, 100, \ldots, 500$ in rows, where ℓ is the number of nearest neighbors. The average annual yield of the nearest-neighbor portfolio is 35%. Comparing Tables 2.4 and 2.5, one can conclude that the nearest neighbor strategy is more robust.

Table 2.5. The average annual yields of the individual experts for the nearest-neighbor strategy.

k ℓ	1	2	3	4	5
50	31%	33%	28%	24%	35%
100	33%	32%	25%	29%	28%
150	38%	33%	26%	32%	27%
200	38%	28%	32%	32%	24%
250	37%	31%	37%	28%	26%
300	41%	35%	35%	30%	29%
350	39%	36%	31%	34%	32%
400	39%	35%	33%	32%	35%
450	39%	34%	34%	35%	37%
500	42%	36%	33%	38%	35%

References

Algoet, P. (1992). Universal schemes for prediction, gambling, and portfolio selection, *Annals of Probability* **20**, pp. 901–941.

Algoet, P. (1994). The strong law of large numbers for sequential decisions under uncertainity, *IEEE Transactions on Information Theory* **40**, pp. 609–634.

Algoet, P. and Cover, T. (1988). Asymptotic optimality and asymptotic equipartition properties of log-optimum investments, *Annals of Probability* **16**, pp. 876–898.

Aumann, R. J. (1977). The St. Petersburg paradox: A discussion of some recent comments, *Journal of Economic Theory* **14**, pp. 443–445.

Barron, A. R. and Cover, T. M. (1988). A bound on the financial value of information, *IEEE Transactions on Information Theory* **34**, pp. 1097–1100.

Bell, R. M. and Cover, T. M. (1980). Competitive optimality of logarithmic investment, *Mathematics of Operations Research* **5**, pp. 161–166.

Bernoulli, D. (1954). Originally published in 1738; translated by L. Sommer. Exposition of a new theory on the measurement of risk, *Econometrica* **22**, pp. 22–36.

Breiman, L. (1961). Optimal gambling systems for favorable games, in *Proc. of the Fourth Berkeley Symposium on Mathematical Statistics and Probability* (University of California Press, Berkeley), pp. 65–78.

Cesa-Bianchi, N. and Lugosi, G. (2006). *Prediction, Learning, and Games* (Cambridge University Press, Cambridge).

Chow, Y. S. (1965). Local convergence of martingales and the law of large numbers, *Annals of Mathematical Statistics* **36**, pp. 552–558.

Chow, Y. S. and Robbins, H. (1961). On sums of independent random variables with infinite moments and "fair" games, in *Proceedings of the National Academy of Sciences* (USA), pp. 330–335.

Cover, T. M. (1984). An algorithm for maximizing expected log investment return, *IEEE Transactions on Information Theory* **30**, pp. 369–373.

Cover, T. M. (1991). Universal portfolios, *Mathematical Finance* **1**, pp. 1–29.

Cover, T. M. and Thomas, J. A. (1991). *Elements of Information Theory* (Wiley, New York).

Csörgő, S. and Simons, G. (1996). A strong law of large numbers for trimmed sums, with applications to generalized St. Petersburg games, *Statistics and Probability Letters* **26**, pp. 65–73.

Durand, D. (1957). Growth stocks and the Petersburg paradox, *The Journal of Finance* **12**, pp. 348–363.

Feller, W. (1945). Note on the law of large numbers and "fair" games, *Annuals of Mathematical Statistics* **16**, pp. 301–304.

Finkelstein, M. and Whitley, R. (1981). Optimal strategies for repeated games, *Advances in Applied Probability* **13**, pp. 415–428.

Gaspero, L. (2006). QuadProg++, C++ porting of the Goldfarb–Idnani dual method, URL http://www.diegm.uniud.it/digaspero/.

Gelencsér, G. and Ottucsák, G. (2006). NYSE data sets at the log-optimal portfolio homepage, URL www.cs.bme.hu/~oti/portfolio.

Goldfarb, D. and Idnani (1983). A numerically stable dual method for solving strictly quadratic programs, *Mathematical Programming* **27**, pp. 1–33.

Györfi, L. and Kevei, P. (2009). St. Petersburg portfolio games, in R. Gavaldà, G. Lugosi, T. Zeugmann and S. Zilles (eds.), *Proceedings of Algorithmic Learning Theory 2009, Lecture Notes in Artificial Intelligence 5809* (Springer-Verlag, Berlin, Heidelberg), pp. 83–96.

Györfi, L. and Kevei, P. (2011). On the rate of convergence of St. Petersburg games, *Periodica Mathematica Hungarica* **62**, pp. 13–37.

Györfi, L., Kohler, M., Krzyżak, A. and Walk, H. (2002). *A Distribution-Free Theory of Nonparametric Regression* (Springer, New York).

Györfi, L., Lugosi, G. and Udina, F. (2006). Nonparametric kernel based sequential investment strategies, *Mathematical Finance* **16**, pp. 337–357.

Györfi, L. and Schäfer, D. (2003). Nonparametric prediction, in J. A. K. Suykens, G. Horváth, S. Basu, C. Micchelli and J. Vandevalle (eds.), *Advances in Learning Theory: Methods, Models and Applications* (IOS Press, NATO Science Series), pp. 339–354.

Györfi, L., Udina, F. and Walk, H. (2008). Nonparametric nearest-neighbor-based empirical portfolio selection strategies, *Statistics and Decisions* **22**, pp. 145–157.

Györfi, L., Urbán, A. and Vajda, I. (2007). Kernel-based semi-log-optimal empirical portfolio selection strategies, *International Journal of Theoretical and Applied Finance* **10**, 5, pp. 505–516.

Haigh, J. (1999). *Taking Chances* (Oxford University Press, Oxford).

Kelly, J. L. (1956). A new interpretation of information rate, *Bell System Technical Journal* **35**, pp. 917–926.

Kenneth, A. J. (1974). The use of unbounded utility functions in expected-utility maximization: Response, *Quarterly Journal of Economics* **88**, pp. 136–138.

Latané, H. A. (1959). Criteria for choice among risky ventures, *Journal of Political Economy* **38**, pp. 145–155.

Luenberger, D. G. (1998). *Investment Science* (Oxford University Press, Oxford).

Markowitz, H. (1952). Portfolio selection, *Journal of Finance* **7**, 1, pp. 77–91.

Markowitz, H. (1976). Investment for the long run: New evidence for an old rule, *Journal of Finance* **31**, 5, pp. 1273–1286.

Martin, R. (2004). The St. Petersburg paradox, in E. N. Zalta (ed.), *The Stanford Encyclopedia of Philosophy (Fall 2004 Edition)* (Stanford University, Stanford, California).

Menger, K. (1934). Das Unsicherheitsmoment in der Wertlehre Betrachtungen im Anschlußan das sogenannte Petersburger Spiel, *Zeitschrift für Nationalökonomie* **5**, pp. 459–485.

Móri, T. F. (1982a). Asymptotic properties of empiricial strategy in favourable stochastic games, in *Proceedings Colloquia Mathematica Societatis János Bolyai 36. Limit Theorems in Probability and Statistics* (North-Holland), pp. 777–790.

Móri, T. F. (1982b). On favourable stochastic games, *Annales Universitatis Scientiarum Budapest. de Rolando Eötvös Nominatae, Sectio Computatorica* **3**, pp. 99–103.

Móri, T. F. (1984). I-divergence geometry of distributions and stochastic games, in *Proceedings of the 3rd Pannonian Symposium on Mathematical Statistics* (North-Holland, Reidel, Dordrecht), pp. 231–238.

Móri, T. F. (1986). Is the empirical strategy optimal? *Statistics and Decisions* **4**, pp. 45–60.

Móri, T. F. and Székely, G. J. (1982). How to win if you can? in *Proceedings Colloquia Mathematica Societatis János Bolyai 36. Limit Theorems in Probability and Statistics*, pp. 791–806.

Morvai, G. (1991). Empirical log-optimal portfolio selection, *Problems of Control and Information Theory* **20**, 6, pp. 453–463.

Morvai, G. (1992). Portfolio choice based on the empirical distribution, *Kybernetika* **28**, 6, pp. 484–493.

Pulley, L. B. (1983). Mean-variance approximation to expected logarithmic utility, *Operations Research* **IT-40**, pp. 685–696.

Rieger, M. O. and Wang, M. (2006). Cumulative prospect theory and the St. Petersburg paradox, *Economic Theory* **28**, pp. 665–679.

Roll, R. (1973). Evidence on the "growth-optimum" model, *The Journal of Finance* **28**, pp. 551–566.

Samuelson, P. A. (1960). The St. Petersburg paradox as a divergent double limit, *International Economic Review* **1**, pp. 31–37.

Singer, Y. (1997). Switching portfolios, *International Journal of Neural Systems* **8**, pp. 445–455.

Spellucci, P. (1999). Donlp2, C Version 7/99. netlib, URL http://www.netlib.org/.

Stout, W. F. (1974). *Almost sure convergence* (Academic Press, New York).

Vajda, I. (2006). Analysis of semi-log-optimal investment strategies, in M. Huskova and M. Janzura (eds.), *Prague Stochastics 2006* (MATFYZ-PRESS, Prague).

Chapter 3

Log-Optimal Portfolio-Selection Strategies with Proportional Transaction Costs

László Györfi

Department of Computer Science and Information Theory,
Budapest University of Technology and Economics,
H-1117, Magyar tudósok körútja 2, Budapest, Hungary.
gyorfi@cs.bme.hu

Harro Walk

Institute of Stochastics and Applications, Universität Stuttgart,
Pfaffenwaldring 57, 70569 Stuttgart, Germany.
walk@mathematik.uni-stuttgart.de

Discrete-time, growth-optimal investment in stock markets with proportional transaction costs is considered. The market process is a sequence of daily relative prices (called returns) and is modeled by a first-order Markov process. Assuming that the distribution of the market process is known, we show sequential investment strategies such that, in the long run, the growth rate on trajectories achieves the maximum, with probability 1. Investments with consumption and with fixed transaction cost where the cost depends on the number of the shares involved in the transaction are also analyzed.

3.1. Introduction

The purpose of this chapter is to investigate sequential investment strategies for financial markets, such that the strategies are allowed to use information collected from the market's past and determine, at the beginning of a trading period, a portfolio; that is, a way to distribute their current capital among the available assets. The goal of the investor is to maximize his wealth in the long run. If there is no transaction cost, the only assumption used in the mathematical analysis is that the daily price relatives form a stationary and ergodic process. Under this assumption, the best strategy (called the log-optimum strategy) can be constructed in full knowledge of the distribution of the entire process, see [Algoet and Cover (1988)]. Moreover, [Györfi and Schäfer (2003)], [Györfi et al. (2006)] and [Györfi et al. (2008)] constructed empirical (data-driven), growth-optimum strategies in the case of unknown distributions. The empirical results show that the performance of these empirical-investment strategies measured on past NYSE data is solid, and sometimes even spectacular.

The problem of optimal investment with proportional transaction cost has been essentially formulated and studied in continuous time only (cf. [Akien et al. (2001)], [Davis and Norman (1990)], [Eastham and Hastings (1988)], [Korn (1998)], [Morton and Pliska (1995)], [Palczewski and Stettner (2006)], [Pliska and Suzuki (2004)], [Shreve et al. (1991)], [Shreve and Soner (1994)], [Taksar et al. (1988)]]).

There are few papers dealing with growth-optimal investment with transaction costs in a discrete-time setting. [Iyengar and Cover (2000)] formulated the problem of horse race markets, where, in every market period, one of the assets has a positive pay off and all the others pay nothing. Their model included proportional transaction costs and they used a long-run, expected average-reward criterion. There are also results for more general markets. [Sass and Schäl (2010)] investigated the numéraire portfolio in the context of bond and stock as assets. [Iyengar (2002, 2005)] investigated growth-optimal investment with several assets assuming an independent and identically distributed (i.i.d.) sequence of asset returns. [Bobryk and Stettner (1999)] considered the case of portfolio selection with consumption, when there are two assets; a bond and a stock. Furthermore, long-run expected discounted reward and i.i.d asset returns were assumed. In the case of a discrete-time and non-i.i.d. market process, [Schäfer (2002)] considered the maximization of the long-run, expected average growth rate with several assets and proportional transaction costs, when the asset returns follow a stationary Markov process. [Györfi and Vajda (2008)] extended the expected growth optimality mentioned above to the almost sure growth optimality.

In this chapter we study the problem of discrete-time, growth-optimal investment in stock markets with proportional, fixed transactions costs and consumption. In Section 3.2 the mathematical setup is introduced. Section 3.3 shows the empirical simulated results of two heuristic algorithms using NYSE data. If the market process is a first-order Markov process and the distribution of the market process is known, then we show in Section 3.4 and Section 3.6 (Proofs) simple sequential investment strategies such that, in the long run, the growth rate on trajectories achieves the maximum with probability 1. Finally, Section 3.5 studies the portfolio selection strategies with consumption and fixed transaction cost.

3.2. Mathematical Setup: Investment with Proportional Transaction Cost

Consider a market consisting of d assets. The evolution of the market in time is represented by a sequence of market vectors $\mathbf{s}_1, \mathbf{s}_2, \ldots \in \mathbb{R}_+^d$, where

$$\mathbf{s}_i = (s_i^{(1)}, \ldots, s_i^{(d)}),$$

such that the j-th component $s_i^{(j)}$ of $\mathbf{s}_i$ denotes the price of the j-th asset at the end of the i-th trading period. ($s_0^{(j)} = 1$.)

In order to apply the usual prediction techniques for time series analysis, one has to transform the sequence $\{\mathbf{s}_i\}$ into a sequence of return vectors $\{\mathbf{x}_i\}$ as follows:

$$\mathbf{x}_i = (x_i^{(1)}, \ldots, x_i^{(d)})$$

such that

$$x_i^{(j)} = \frac{s_i^{(j)}}{s_{i-1}^{(j)}}.$$

Thus, the j-th component $x_i^{(j)}$ of the return vector $\mathbf{x}_i$ denotes the amount obtained at the end of the i-th trading period after investing a unit capital in the j-th asset.

The investor is allowed to diversify his capital at the beginning of each trading period according to a portfolio vector $\mathbf{b} = (b^{(1)}, \ldots b^{(d)})^T$. The j-th component $b^{(j)}$ of $\mathbf{b}$ denotes the proportion of the investor's capital invested in asset j. Throughout the chapter, we assume that the portfolio vector $\mathbf{b}$ has nonnegative components with $\sum_{j=1}^d b^{(j)} = 1$. The fact that $\sum_{j=1}^d b^{(j)} = 1$ means that the investment strategy is self-financing and consumption of capital is excluded (except Section 3.5). The nonnegativity of the components of $\mathbf{b}$ means that short selling and buying stocks on margin are not permitted. To make the analysis feasible, some simplifying assumptions are used that need to be taken into account. We assume that assets are arbitrarily divisible and all assets are available in unbounded quantities at the current price in any given trading period. We also assume that the behavior of the market is not affected by the actions of the investor using the strategies under investigation.

For $j \leq i$ we abbreviate the array of return vectors $(\mathbf{x}_j, \ldots, \mathbf{x}_i)$ by $\mathbf{x}_j^i$. Denote by Δ_d the simplex of all vectors $\mathbf{b} \in \mathbb{R}_+^d$ with nonnegative components summing up to one. An *investment strategy* is a sequence $\mathbf{B}$ of functions

$$\mathbf{b}_i : \left(\mathbb{R}_+^d \right)^{i-1} \to \Delta_d, \qquad i = 1, 2, \ldots$$

so that $\mathbf{b}_i(\mathbf{x}_1^{i-1})$ denotes the portfolio vector chosen by the investor on the i-th trading period, upon observing the past behavior of the market. We write $\mathbf{b}(\mathbf{x}_1^{i-1}) = \mathbf{b}_i(\mathbf{x}_1^{i-1})$ to simplify the notation.

In this section, our presentation of the transaction cost problem utilizes the formulation in [Kalai and Blum (1999)], [Schäfer (2002)] and [Györfi and Vajda (2008)]. Let S_n denote the gross wealth at the end of trading period n, $n = 0, 1, 2, \cdots$, where, without loss of generality, let the investor's initial capital S_0 be \$1, while N_n stands for the net wealth at the end of trading period n. Using the above notations, for the trading period n, the net wealth N_{n-1} can be invested according to the portfolio $\mathbf{b}_n$, therefore, the gross wealth S_n at the end of trading period n is

$$S_n = N_{n-1} \sum_{j=1}^{d} b_n^{(j)} x_n^{(j)} = N_{n-1} \langle \mathbf{b}_n , \mathbf{x}_n \rangle,$$

where $\langle \cdot , \cdot \rangle$ denotes inner product.

At the beginning of a new market day $n + 1$, the investor sets up his new portfolio, i.e., buys/sells stocks according to the actual portfolio vector $\mathbf{b}_{n+1}$. During this rearrangement, he has to pay transaction costs, therefore, at the beginning of a new market day $n + 1$, the net wealth N_n in the portfolio $\mathbf{b}_{n+1}$ is less than S_n.

The rates of proportional transaction costs (commission factors) levied on one asset are denoted by $0 < c_s < 1$ and $0 < c_p < 1$, i.e., the sale of \$1 worth of asset i nets only $1 - c_s$ dollars, and similarly we take into account the purchase of an asset such that the purchase of one dollar's worth of asset i costs an extra c_p dollars. We consider the special case when the rate of costs are constant over the assets.

Let us calculate the transaction cost to be paid when selecting the portfolio $\mathbf{b}_{n+1}$. Before rearranging the capitals, at the j-th asset there are $b_n^{(j)} x_n^{(j)} N_{n-1}$ dollars, while, after rearranging, we need $b_{n+1}^{(j)} N_n$ dollars. If $b_n^{(j)} x_n^{(j)} N_{n-1} \geq b_{n+1}^{(j)} N_n$ then we have to sell and the transaction cost at the j-th asset is

$$c_s \left(b_n^{(j)} x_n^{(j)} N_{n-1} - b_{n+1}^{(j)} N_n \right),$$

otherwise we have to buy and the transaction cost at the j-th asset is

$$c_p \left(b_{n+1}^{(j)} N_n - b_n^{(j)} x_n^{(j)} N_{n-1} \right).$$

Let x^+ denote the positive part of x. Thus, the gross wealth S_n decomposes to the sum of the net wealth and cost in the following self-financing way:

$$N_n = S_n - \sum_{j=1}^{d} c_s \left(b_n^{(j)} x_n^{(j)} N_{n-1} - b_{n+1}^{(j)} N_n \right)^+$$
$$- \sum_{j=1}^{d} c_p \left(b_{n+1}^{(j)} N_n - b_n^{(j)} x_n^{(j)} N_{n-1} \right)^+,$$

or, equivalently,

$$S_n = N_n + c_s \sum_{j=1}^{d} \left(b_n^{(j)} x_n^{(j)} N_{n-1} - b_{n+1}^{(j)} N_n \right)^+$$
$$+ c_p \sum_{j=1}^{d} \left(b_{n+1}^{(j)} N_n - b_n^{(j)} x_n^{(j)} N_{n-1} \right)^+.$$

Dividing both sides by S_n and introducing the ratio

$$w_n = \frac{N_n}{S_n},$$

$0 < w_n < 1$, we obtain

$$1 = w_n + c_s \sum_{j=1}^{d} \left(\frac{b_n^{(j)} x_n^{(j)}}{\langle \mathbf{b}_n , \mathbf{x}_n \rangle} - b_{n+1}^{(j)} w_n \right)^+$$
$$+ c_p \sum_{j=1}^{d} \left(b_{n+1}^{(j)} w_n - \frac{b_n^{(j)} x_n^{(j)}}{\langle \mathbf{b}_n , \mathbf{x}_n \rangle} \right)^+. \tag{3.1}$$

For a given previous return vector $\mathbf{x}_n$ and portfolio vector $\mathbf{b}_n$, there is a portfolio vector $\tilde{\mathbf{b}}_{n+1} = \tilde{\mathbf{b}}_{n+1}(\mathbf{b}_n, \mathbf{x}_n)$ for which there is no trading:

$$\tilde{b}_{n+1}^{(j)} = \frac{b_n^{(j)} x_n^{(j)}}{\langle \mathbf{b}_n , \mathbf{x}_n \rangle} \tag{3.2}$$

such that there is no transaction cost, i.e., $w_n = 1$.

For arbitrary portfolio vectors $\mathbf{b}_n$, $\mathbf{b}_{n+1}$, and return vector $\mathbf{x}_n$ there exist unique cost factors $w_n \in [0, 1)$. The value of cost factor w_n at day n

is determined by portfolio vectors $\mathbf{b}_n$ and $\mathbf{b}_{n+1}$, as well as by return vector $\mathbf{x}_n$, i.e.,

$$w_n = w(\mathbf{b}_n, \mathbf{b}_{n+1}, \mathbf{x}_n),$$

for some function w. If we want to rearrange our portfolio substantially, then our net wealth decreases more considerably; however, it remains positive. Note also that the cost does not restrict the set of new portfolio vectors, i.e., the optimization algorithm searches for optimal vector $\mathbf{b}_{n+1}$ within the whole simplex Δ_d. The value of the cost factor ranges between

$$\frac{1 - c_s}{1 + c_p} \leq w_n \leq 1.$$

For the sake of simplicity, we consider the special case of $c \triangleq c_s = c_p$. Then,

$$c_s \left(b_n^{(j)} x_n^{(j)} N_{n-1} - b_{n+1}^{(j)} N_n \right)^+ + c_p \left(b_{n+1}^{(j)} N_n - b_n^{(j)} x_n^{(j)} N_{n-1} \right)^+$$
$$= c \left| b_n^{(j)} x_n^{(j)} N_{n-1} - b_{n+1}^{(j)} N_n \right|.$$

Starting with an initial wealth $S_0 = 1$ and $w_0 = 1$, wealth S_n at the closing time of the n-th market day becomes

$$S_n = N_{n-1} \langle \mathbf{b}_n, \mathbf{x}_n \rangle$$
$$= w_{n-1} S_{n-1} \langle \mathbf{b}_n, \mathbf{x}_n \rangle$$
$$= \prod_{i=1}^{n} [w(\mathbf{b}_{i-1}, \mathbf{b}_i, \mathbf{x}_{i-1}) \langle \mathbf{b}_i, \mathbf{x}_i \rangle].$$

Introduce the notation

$$g(\mathbf{b}_{i-1}, \mathbf{b}_i, \mathbf{x}_{i-1}, \mathbf{x}_i) = \ln(w(\mathbf{b}_{i-1}, \mathbf{b}_i, \mathbf{x}_{i-1}) \langle \mathbf{b}_i, \mathbf{x}_i \rangle),$$

then, the average growth rate becomes

$$\frac{1}{n} \ln S_n = \frac{1}{n} \sum_{i=1}^{n} \ln(w(\mathbf{b}_{i-1}, \mathbf{b}_i, \mathbf{x}_{i-1}) \langle \mathbf{b}_i, \mathbf{x}_i \rangle)$$
$$= \frac{1}{n} \sum_{i=1}^{n} g(\mathbf{b}_{i-1}, \mathbf{b}_i, \mathbf{x}_{i-1}, \mathbf{x}_i). \tag{3.3}$$

Our aim is to maximize this average growth rate.

In what follows, $\mathbf{x}_i$ will be a random variable and is denoted by $\mathbf{X}_i$, and we assume the following conditions:

(i) $\{\mathbf{X}_i\}$ is a homogeneous and first-order Markov process,

(ii) the Markov kernel is continuous, which means that, for $\mu(B|\mathbf{x})$ being the Markov kernel defined by

$$\mu(B|\mathbf{x}) \triangleq \mathbb{P}\{\mathbf{X}_2 \in B \mid \mathbf{X}_1 = \mathbf{x}\},$$

we assume that the Markov kernel is continuous in total variation, i.e.,

$$V(\mathbf{x}, \mathbf{x}') \triangleq \sup_{B \in \mathcal{B}} |\mu(B|\mathbf{x}) - \mu(B|\mathbf{x}')| \to 0$$

if $\mathbf{x}' \to \mathbf{x}$ such that $\mathcal{B}$ denotes the family of Borel σ-algebra, further

$$V(\mathbf{x}, \mathbf{x}') < 1 \text{ for all } \mathbf{x}, \mathbf{x}',$$

(iii) and there exist $0 < a_1 < 1 < a_2 < \infty$, such that $a_1 \le X^{(j)} \le a_2$ for all $j = 1, \dots, d$.

We note that Conditions (ii) and (iii) imply uniform continuity of V and, thus,

$$\max_{\mathbf{x}, \mathbf{x}'} V(\mathbf{x}, \mathbf{x}') < 1. \tag{3.4}$$

For the usual stock market daily data, Condition (iii) is satisfied with $a_1 = 0.7$ and with $a_2 = 1.2$ (cf. [Fernholz (2000)]).

In the realistic case that the state space of the Markov process $(\mathbf{X}_n)$ is a finite set D of rational vectors (components being quotients of integer-valued dollar-amounts) containing $\mathbf{e} = (1, \dots, 1)$, the second part of condition (ii) is fulfilled under the plausible assumption $\mu(\{\mathbf{e}\}|\mathbf{x}) > 0$ for all $\mathbf{x} \in D$. Another example for a finite-state Markov process is when one rounds down the components of $\mathbf{x}$ to a grid applying, for example, a grid size of 0.00001.

Let us use the decomposition

$$\frac{1}{n} \ln S_n = I_n + J_n, \tag{3.5}$$

where I_n is

$$\frac{1}{n} \sum_{i=1}^{n} (g(\mathbf{b}_{i-1}, \mathbf{b}_i, \mathbf{X}_{i-1}, \mathbf{X}_i) - \mathbb{E}\{g(\mathbf{b}_{i-1}, \mathbf{b}_i, \mathbf{X}_{i-1}, \mathbf{X}_i)|\mathbf{X}_1^{i-1}\})$$

and

$$J_n = \frac{1}{n} \sum_{i=1}^{n} \mathbb{E}\{g(\mathbf{b}_{i-1}, \mathbf{b}_i, \mathbf{X}_{i-1}, \mathbf{X}_i)|\mathbf{X}_1^{i-1}\}.$$

I_n is an average of martingale differences. Under Condition (iii), the random variable $g(\mathbf{b}_{i-1}, \mathbf{b}_i, \mathbf{X}_{i-1}, \mathbf{X}_i)$ is bounded, therefore I_n is an average of bounded martingale differences, which converges to 0 almost surely, since, according to the Chow Theorem (cf. Theorem 3.3.1 in [Stout (1974)]),

$$\sum_{i=1}^{\infty} \frac{\mathbb{E}\{g(\mathbf{b}_{i-1}, \mathbf{b}_i, \mathbf{X}_{i-1}, \mathbf{X}_i)^2\}}{i^2} < \infty$$

implies that

$$I_n \to 0 \qquad \text{almost surely.}$$

Thus, the asymptotic maximization of the average growth rate $\frac{1}{n} \ln S_n$ is equivalent to the maximization of J_n.

Under Condition (i), we have that

$$
\begin{aligned}
& \mathbb{E}\{g(\mathbf{b}_{i-1}, \mathbf{b}_i, \mathbf{X}_{i-1}, \mathbf{X}_i)|\mathbf{X}_1^{i-1}\} \\
&= \mathbb{E}\{\ln(w(\mathbf{b}_{i-1}, \mathbf{b}_i, \mathbf{X}_{i-1}) \langle \mathbf{b}_i , \mathbf{X}_i\rangle)|\mathbf{X}_1^{i-1}\} \\
&= \ln w(\mathbf{b}_{i-1}, \mathbf{b}_i, \mathbf{X}_{i-1}) + \mathbb{E}\{\ln \langle \mathbf{b}_i , \mathbf{X}_i\rangle |\mathbf{X}_1^{i-1}\} \\
&= \ln w(\mathbf{b}_{i-1}, \mathbf{b}_i, \mathbf{X}_{i-1}) + \mathbb{E}\{\ln \langle \mathbf{b}_i , \mathbf{X}_i\rangle |\mathbf{b}_i, \mathbf{X}_{i-1}\} \\
&\triangleq v(\mathbf{b}_{i-1}, \mathbf{b}_i, \mathbf{X}_{i-1}).
\end{aligned}
$$

Therefore, the maximization of the average growth rate $\frac{1}{n} \ln S_n$ is asymptotically equivalent to the maximization of

$$J_n = \frac{1}{n} \sum_{i=1}^{n} v(\mathbf{b}_{i-1}, \mathbf{b}_i, \mathbf{X}_{i-1}). \tag{3.6}$$

The terms in the average J_n have a memory, which transforms the problem into a dynamic-programming setup (cf. [Merhav *et al.* (2002)]).

3.3. Experiments on Heuristic Algorithms

In this section we experimentally study two heuristic algorithms, which performed well without transaction cost (cf. Chapter 2 of this volume).

Algorithm 1. For transaction cost, one may apply the log-optimal portfolio

$$\mathbf{b}_n^*(\mathbf{X}_{n-1}) = \arg\max_{\mathbf{b}(\cdot)} \mathbb{E}\{\ln \langle \mathbf{b}(\mathbf{X}_{n-1}), \mathbf{X}_n\rangle \mid \mathbf{X}_{n-1}\}$$

or its empirical approximation. For example, we may apply the kernel-based log-optimal portfolio selection introduced by [Györfi *et al.* (2006)] as follows. Define an infinite array of experts $\mathbf{B}^{(\ell)} = \{\mathbf{b}^{(\ell)}(\cdot)\}$, where ℓ is a positive integer. For a fixed positive integer ℓ, choose the radius $r_\ell > 0$ such that

$$\lim_{\ell \to \infty} r_\ell = 0.$$

Then, for $n > 1$, define the expert $\mathbf{b}^{(\ell)}$ as follows. Put

$$\mathbf{b}_n^{(\ell)} = \arg\max_{\mathbf{b} \in \Delta_d} \sum_{\{i < n: \|\mathbf{x}_{i-1} - \mathbf{x}_{n-1}\| \leq r_\ell\}} \ln \langle \mathbf{b}, \mathbf{x}_i \rangle, \qquad (3.7)$$

if the sum is non-void, and $\mathbf{b}_0 = (1/d, \dots, 1/d)$ otherwise, where $\|\cdot\|$ denotes the Euclidean norm.

Similarly to Chapter 2 of this volume, these experts are aggregated (mixed) as follows. Let $\{q_\ell\}$ be a probability distribution over the set of all positive integers ℓ such that for all ℓ, $q_\ell > 0$. Consider two types of aggregations:

- Here, the initial capital $S_0 = 1$ is distributed among the experts according to the distribution $\{q_\ell\}$, and the experts make the portfolio selection and pay for transaction cost individually. If $S_n(\mathbf{B}^{(\ell)})$ is the capital accumulated by the elementary strategy $\mathbf{B}^{(\ell)}$ after n periods when starting with an initial capital $S_0 = 1$, then, after period n, the investor's aggregated wealth is

$$S_n = \sum_\ell q_\ell S_n(\mathbf{B}^{(\ell)}). \qquad (3.8)$$

- Here, $S_n(\mathbf{B}^{(\ell)})$ is again the capital accumulated by the elementary strategy $\mathbf{B}^{(\ell)}$ after n periods when starting with an initial capital $S_0 = 1$, but it is virtual figure, i.e., the experts make no trading, its wealth is just the base of aggregation. Then, after period n, the investor's aggregated portfolio becomes

$$\mathbf{b}_n = \frac{\sum_\ell q_\ell S_{n-1}(\mathbf{B}^{(\ell)}) \mathbf{b}_n^{(\ell)}}{\sum_\ell q_\ell S_{n-1}(\mathbf{B}^{(\ell)})}. \qquad (3.9)$$

Moreover, the investor's capital is

$$S_n = S_{n-1} \langle \mathbf{b}_n, \mathbf{x}_n \rangle w(\mathbf{b}_{n-1}, \mathbf{b}_n, \mathbf{x}_{n-1}),$$

so only the aggregated portfolio pays for the transaction cost.

In Chapter 2 of this volume we proved that, without transaction cost, the two aggregations are equivalent. However, when including transaction cost, the aggregation (3.9) is much better.

Algorithm 2. We may introduce a suboptimal algorithm, called a *naïve portfolio*, by a one-step optimization as follows. Put $\mathbf{b}_1 = \{1/d, \ldots, 1/d\}$ and, for $n \geq 1$,

$$\mathbf{b}_n^{(\ell)} = \arg\max_{\mathbf{b} \in \Delta_d} \sum_{\{i < n : \|\mathbf{x}_{i-1} - \mathbf{x}_{n-1}\| \leq r_\ell\}} \left(\ln \langle \mathbf{b}, \mathbf{x}_i \rangle + \ln w(\mathbf{b}_{n-1}, \mathbf{b}, \mathbf{x}_{n-1}) \right),$$

(3.10)

if the sum is non-void, and $\mathbf{b}_0 = (1/d, \ldots, 1/d)$ otherwise. These elementary portfolios are mixed as in (3.8) or (3.9). Obviously, this portfolio has no global optimality property.

Next, we present some numerical results for transaction cost obtained by applying the kernel-based semi-log-optimal algorithm to the 19 assets of the second NYSE data set as in Chapter 2 of this volume. We take a finite set of of experts of size L. In the experiment, we selected $L = 10$. Choose the uniform distribution $q_\ell = 1/L$ over the experts in use, and the radius

$$r_\ell^2 = 0.0002 \cdot d(1 + \ell/10), \qquad \text{for } \ell = 1, \ldots, L .$$

Table 3.1 summarizes the average annual yield achieved by each expert in the last period when investing one unit for the kernel-based log-optimal portfolio. Experts are indexed by $\ell = 1 \ldots 10$ in rows. The second column contains the average annual yields of experts for the kernel-based log-optimal portfolio if there is no transaction cost, and, in this case, the results of the two aggregations are the same; 35%. Note that, out of the 19 assets, MORRIS had the best average annual yield, 20%, so, for no transaction cost, with the kernel-based log-optimal portfolio, we have a spectacular improvement. The third and fourth columns contain the average annual yields of experts for a kernel-based log-optimal portfolio if the commission factor is $c = 0.0015$. Notice that the growth rate of the Algorithm 1 is negative, and the growth rate of the Algorithm 2 is also poor it is less than the growth rate of the best asset, and the results of aggregations are different.

In Table 3.2 we have obtained similar results for the nearest-neighbor strategy, where ℓ is the number of nearest neighbors. As we mentioned in Chapter 2 of this volume, the time-varying portfolio is very undiversified such that the subset of assets with non-zero weight changes from time to time, which makes the problem of transaction cost challenging. Moreover, the better the nearest neighbor strategy is without transaction cost, the worse it is with transaction cost, and the main reasoning of this fact is that, for the good time-varying portfolio, the portfolio vector component is highly fluctuating, and, so, the proper handling of the transaction cost is still an open question and an important direction of future research.

Table 3.1. The average annual yields of the individual experts for kernel strategy and of the aggregations with $c = 0.0015$.

ℓ	$c = 0$	Algorithm 1	Algorithm 2
1	31%	-22%	18%
2	34%	-22%	10%
3	35%	-24%	9 %
4	35%	-23%	14%
5	34%	-21%	13%
6	35%	-19%	13%
7	33%	-20%	12%
8	34%	-18%	8 %
9	37%	-17%	6 %
10	34%	-18%	11%
Wealth Agg. (3.8)	35%	**-19%**	**13%**
Portfolio Agg. (3.9)	35%	**-15%**	**17%**

Table 3.2. The average annual yields of the individual experts for the nearest-neighbor strategy and of the aggregations with $c = 0.0015$.

ℓ	$c = 0$	Algorithm 1	Algorithm 2
50	31%	-35%	-14%
100	33%	-33%	3%
150	38%	-29%	3%
200	38%	-28%	9%
250	37%	-28%	9%
300	41%	-26%	7%
350	39%	-26%	9%
400	39%	-26%	10%
450	39%	-25%	14%
500	42%	-23%	14%
Wealth Agg. (3.8)	39%	**-25%**	**11%**
Portfolio Agg. (3.9)	39%	**-23%**	**11%**

3.4. Growth-Optimal Portfolio Selection Algorithms

Bellman-type optimality equations are essential tools in the definition and investigation of portfolio-selection algorithms under the presence of transaction costs. First, we present an informal and heuristic introduction to them in our context of portfolio selection. Later on, a rigorous treatment will be given.

Let us start with a finite-horizon problem concerning J_N defined by (3.6). For a fixed integer $N > 0$, maximize

$$\mathbb{E}\{N \cdot J_N \mid \mathbf{b}_0 = \mathbf{b}, \mathbf{X}_0 = \mathbf{x}\} = \mathbb{E}\left\{\sum_{i=1}^{N} v(\mathbf{b}_{i-1}, \mathbf{b}_i, \mathbf{X}_{i-1}) \mid \mathbf{b}_0 = \mathbf{b}, \mathbf{X}_0 = \mathbf{x}\right\}$$

by a suitable choice of $\mathbf{b}_1, \ldots, \mathbf{b}_N$. For general problems of dynamic-programming (dynamic optimization), [Bellman (1957)] on page 89 formulates his famous principle of optimality as follows: "An optimality policy has the property that whatever the initial state and initial decisions are, the remaining decisions must constitute an optimal policy with regard to the state resulting from the first decision."

By this principle, which, for stochastic models, is not so obvious as it seems (cf. pp. 14, 15 in [Hinderer (1970)]), one can show the following. Let the functions $G_0, G_1, \ldots, G_N$ on $\Delta_d \times [a_1, a_2]^d$ be defined by the so-called dynamic-programming equations (optimality equations, Bellman equations)

$$G_N(\mathbf{b}, \mathbf{x}) \triangleq 0,$$
$$G_n(\mathbf{b}, \mathbf{x}) \triangleq \max_{\mathbf{b}'} [v(\mathbf{b}, \mathbf{b}', \mathbf{x}) + \mathbb{E}\{G_{n+1}(\mathbf{b}', \mathbf{X}_2) \mid \mathbf{X}_1 = \mathbf{x}\}]$$

$(n = N - 1, N - 2, \ldots, 0)$ with maximizer $\mathbf{b}'_n = g_n(\mathbf{b}, \mathbf{x})$. Setting

$$F^n \triangleq G_{N-n}$$

$(n = 0, 1, \ldots, N)$, one can write these backward equations in the forward form

$$F^0(\mathbf{b}, \mathbf{x}) \triangleq 0,$$
$$F^n(\mathbf{b}, \mathbf{x}) \triangleq \max_{\mathbf{b}'} [v(\mathbf{b}, \mathbf{b}', \mathbf{x}) + \mathbb{E}\{F^{n-1}(\mathbf{b}', \mathbf{X}_2) \mid \mathbf{X}_1 = \mathbf{x}\}] \quad (3.11)$$

$(n = 1, 2, \ldots, N)$ with maximizer $f_n(\mathbf{b}, \mathbf{x}) = g_{N-n}(\mathbf{b}, \mathbf{x})$. Then, the choices $\mathbf{b}_n = f_n(\mathbf{b}_{n-1}, \mathbf{X}_{n-1})$ are optimal.

For the situations which are favorite for the investor, one has $F^n(\mathbf{b}, \mathbf{x}) \to \infty$ as $n \to \infty$, which does not allow distinguishing between the qualities of competing choice sequences in the infinite-horizon case. If one considers (3.11) as a *value iteration* formula, then the underlying Bellman-type equation

$$F^\infty(\mathbf{b}, \mathbf{x}) = \max_{\mathbf{b}'} \{v(\mathbf{b}, \mathbf{b}', \mathbf{x}) + \mathbb{E}\{F^\infty(\mathbf{b}', \mathbf{X}_2) \mid \mathbf{X}_1 = \mathbf{x}\}\}$$

has, roughly speaking, the degenerate solution $F^\infty = \infty$. Therefore, one uses a discount factor $0 < \delta < 1$ and arrives at the discounted Bellman-equation

$$F_\delta(\mathbf{b}, \mathbf{x}) = \max_{\mathbf{b}'} \{v(\mathbf{b}, \mathbf{b}', \mathbf{x}) + (1 - \delta)\mathbb{E}\{F_\delta(\mathbf{b}', \mathbf{X}_2) \mid \mathbf{X}_1 = \mathbf{x}\}\}. \quad (3.12)$$

Its solution allows us to solve the discounted problem, maximizing

$$\mathbb{E}\left\{\sum_{i=1}^{\infty}(1 - \delta)^i v(\mathbf{b}_{i-1}, \mathbf{b}_i, \mathbf{X}_{i-1}) \mid \mathbf{b}_0 = \mathbf{b}, \mathbf{X}_0 = \mathbf{x}\right\}$$

$$= \sum_{i=1}^{\infty}(1 - \delta)^i \mathbb{E}\left\{v(\mathbf{b}_{i-1}, \mathbf{b}_i, \mathbf{X}_{i-1}) \mid \mathbf{b}_0 = \mathbf{b}, \mathbf{X}_0 = \mathbf{x}\right\}.$$

The classic Hardy–Littlewood theorem (see, e.g., Theorem 95, together with Theorem 55 in [Hardy (1949)]) states that, for a real-valued, bounded sequence a_n, $n = 1, 2, \dots$,

$$\lim_{\delta \downarrow 0} \delta \sum_{i=0}^{\infty}(1 - \delta)^i a_i$$

exists if and only if

$$\lim_{n \to \infty} \frac{1}{n} \sum_{i=0}^{n-1} a_i$$

exists and that, then, the limits are equal. Therefore, for maximizing

$$\lim_{n \to \infty} \frac{1}{n} \sum_{i=1}^{n} \mathbb{E}\left\{v(\mathbf{b}_{i-1}, \mathbf{b}_i, \mathbf{X}_{i-1}) \mid \mathbf{b}_0 = \mathbf{b}, \mathbf{X}_0 = \mathbf{x}\right\},$$

(if it exists), it is important to solve Equation (3.12) for small δ. This principle results in Rule 1 below. Letting $\delta \downarrow 0$, (3.12) with solution F_δ^* leads to the non-discounted Bellman-equation

$$\lambda + F(\mathbf{b}, \mathbf{x}) = \max_{\mathbf{b}'} \{v(\mathbf{b}, \mathbf{b}', \mathbf{x}) + \mathbb{E}\{F(\mathbf{b}', \mathbf{X}_2) \mid \mathbf{X}_1 = \mathbf{x}\}\}. \quad (3.13)$$

The interpretation of (3.11) as a value iteration motivates solving (3.12) and (3.13) also by value iterations $F_{\delta,n}$ (see below) and F_n' with discount factors $\delta_n \downarrow 0$ (see Rule 4). As to the corresponding problems in Markov-control theory we refer to [Hernández-Lerma and Lasserre (1996)].

[Györfi and Vajda (2008)] studied the following two optimal-portfolio selection rules. Let $0 < \delta < 1$ denote a discount factor. Let the discounted Bellman-equation be (3.12). One can show that this discounted Bellman-equation, (3.12), and also the more general Bellman-equation, (3.19) below, have a unique solution (cf. [Schäfer (2002)] and the proof of Proposition 3.1 below). Concerning the discounted Bellman-equation (3.12), the so-called value iteration may result in the solution. For fixed $0 < \delta < 1$ and for $k = 0, 1, \ldots$, put

$$F_{\delta,0} = 0$$

and

$$F_{\delta,k+1}(\mathbf{b}, \mathbf{x}) = \max_{\mathbf{b}'} \left\{ v(\mathbf{b}, \mathbf{b}', \mathbf{x}) + (1 - \delta)\mathbb{E}\{F_{\delta,k}(\mathbf{b}', \mathbf{X}_2) \mid \mathbf{X}_1 = \mathbf{x}\} \right\}.$$

Then Banach's fixed point theorem implies that the value iteration converges uniformly to the unique solution.

Rule 1. [Schäfer (2002)] introduced the following non-stationary rule. Put

$$\bar{\mathbf{b}}_1 = \{1/d, \ldots, 1/d\}$$

and

$$\bar{\mathbf{b}}_{i+1} = \arg\max_{\mathbf{b}'} \left\{ v(\bar{\mathbf{b}}_i, \mathbf{b}', \mathbf{X}_i) + (1 - \delta_i)\mathbb{E}\{F_{\delta_i}(\mathbf{b}', \mathbf{X}_{i+1})|\mathbf{X}_i\} \right\},$$

for $1 \leq i$, where $0 < \delta_i < 1$ is a discount factor such that $\delta_i \downarrow 0$. [Schäfer (2002)] proved that, for the conditions (i), (ii) (in a weakened form) and (iii) and, under some mild conditions on δ_i's for Rule 1, the portfolio $\{\bar{\mathbf{b}}_i\}$ with capital $\bar{S}_n$ is optimal, in the sense that, for any portfolio strategy $\{\mathbf{b}_i\}$ with capital S_n,

$$\liminf_{n\to\infty} \left(\frac{1}{n}\mathbb{E}\{\ln \bar{S}_n\} - \frac{1}{n}\mathbb{E}\{\ln S_n\} \right) \geq 0.$$

[Győrfi and Vajda (2008)] extended this optimality in expectation to pathwise optimality such that, under the same conditions,

$$\liminf_{n\to\infty} \left(\frac{1}{n}\ln \bar{S}_n - \frac{1}{n}\ln S_n \right) \geq 0 \qquad \text{almost surely.}$$

Rule 2. [Győrfi and Vajda (2008)] introduced a portfolio with *stationary* (time-invariant) recursion. For any integer $k \geq 1$, put

$$\mathbf{b}_1^{(k)} = \{1/d, \ldots, 1/d\}$$

and

$$\mathbf{b}_{i+1}^{(k)} = \arg\max_{\mathbf{b}'} \left\{ v(\mathbf{b}_i^{(k)}, \mathbf{b}', \mathbf{X}_i) + (1 - \delta_k)\mathbb{E}\{F_{\delta_k}(\mathbf{b}', \mathbf{X}_{i+1})|\mathbf{X}_i\} \right\},$$

for $i \geq 1$, where $0 < \delta_k < 1$. The portfolio $\mathbf{B}^{(k)} = \{\mathbf{b}_i^{(k)}\}$ is called the portfolio of expert k with capital $S_n(\mathbf{B}^{(k)})$. Choose an arbitrary probability distribution $q_k > 0$ and introduce the combined portfolio with its capital

$$\tilde{S}_n = \sum_{k=1}^{\infty} q_k S_n(\mathbf{B}^{(k)}).$$

[Györfi and Vajda (2008)] proved that, under the above-mentioned conditions, for Rule 2,

$$\lim_{n \to \infty} \left(\frac{1}{n} \ln \bar{S}_n - \frac{1}{n} \ln \tilde{S}_n \right) = 0 \qquad \text{almost surely.}$$

Notice that possibly none of the averaged growth rates $\frac{1}{n} \ln \bar{S}_n$ and $\frac{1}{n} \ln \tilde{S}_n$ are convergent to a constant, since we did not assume the ergodicity of $\{\mathbf{X}_i\}$.

Next, we introduce further portfolio-selection rules. According to Proposition 3.1 below, a solution $(\lambda = W_c^*, F)$ of the (non-discounted) Bellman-equation (3.13) exists, where $W_c^* \in \mathbb{R}$ is unique according to Proposition 3.2 below. W_c^* is the maximum growth rate (see Theorem 3.1 below).

Rule 3. Introduce a stationary rule such that

$$\mathbf{b}_1^* = \{1/d, \dots, 1/d\}$$

and

$$\mathbf{b}_{i+1}^* = \arg\max_{\mathbf{b}'} \{v(\mathbf{b}_i^*, \mathbf{b}', \mathbf{X}_i) + \mathbb{E}\{F(\mathbf{b}', \mathbf{X}_{i+1}) | \mathbf{X}_i\}\}. \tag{3.14}$$

Theorem 3.1. *Under the conditions (i), (ii) and (iii), if S_n^* denotes the wealth at period n using the portfolio $\{\mathbf{b}_n^*\}$, then*

$$\lim_{n \to \infty} \frac{1}{n} \ln S_n^* = W_c^* \qquad \text{almost surely,}$$

while, if S_n denotes the wealth at period n using any other portfolio $\{\mathbf{b}_n\}$, then

$$\limsup_{n \to \infty} \frac{1}{n} \ln S_n \leq W_c^* \qquad \text{almost surely.}$$

Remark 3.1. There is the obvious question of how to ensure that $W_c^* > 0$. Next, we show a simple sufficient condition for $W_c^* > 0$. We prove that, if the best asset has positive growth rate, then $W_c^* > 0$, for any c. Consider the uniform static portfolio (uniform index), i.e., at time $n = 0$ we apply the uniform portfolio and later on there is no trading. It means that the wealth at time n is defined by

$$S_n = S_0 \frac{1}{d} \sum_{j=1}^{d} s_n^{(j)}.$$

Apply the following simple bounds

$$S_0 \frac{1}{d} \max_j s_n^{(j)} \leq S_n \leq S_0 \max_j s_n^{(j)}.$$

These bounds imply that

$$\limsup_{n \to \infty} \frac{1}{n} \ln S_n = \limsup_{n \to \infty} \max_j \frac{1}{n} \ln s_n^{(j)}$$

$$\geq \max_j \limsup_{n \to \infty} \frac{1}{n} \ln s_n^{(j)}$$

$$\triangleq \max_j W^{(j)} > 0.$$

Thus,

$$W_c^* \geq \max_j W^{(j)} > 0.$$

Remark 3.2. For an i.i.d. market process, [Iyengar (2002, 2005)] observed that, even in a discrete-time case there is no trading with positive probability, i.e.,

$$\mathbb{P}\{\tilde{\mathbf{b}}_{n+1}(\mathbf{b}_n^*, \mathbf{X}_n) = \mathbf{b}_{n+1}^*\} > 0,$$

where the no-trading portfolio $\tilde{\mathbf{b}}_{n+1}$ has been defined by (3.2). Moreover, one may obtain an approximately-optimal selection rule, if $\mathbf{b}_{n+1}^*$ is restricted on an appropriate neighborhood of $\tilde{\mathbf{b}}_{n+1}(\mathbf{b}_n^*, \mathbf{X}_n)$.

Remark 3.3. The problem is simpler if the market process is i.i.d. Then, on the one hand v has the form

$$v(\mathbf{b}, \mathbf{b}', \mathbf{x}) = \ln w(\mathbf{b}, \mathbf{b}', \mathbf{x}) + \mathbb{E}\{\ln \langle \mathbf{b}', \mathbf{X}_2 \rangle \,|\, \mathbf{X}_1 = \mathbf{x}\}$$

$$= \ln w(\mathbf{b}, \mathbf{b}', \mathbf{x}) + \mathbb{E}\{\ln \langle \mathbf{b}', \mathbf{X}_2 \rangle\},$$

while the Bellman-equation (3.13) looks like

$$W_c^* + F(\mathbf{b}, \mathbf{x}) = \max_{\mathbf{b}'} \left\{ v(\mathbf{b}, \mathbf{b}', \mathbf{x}) + \mathbb{E}\{F(\mathbf{b}', \mathbf{X}_2) \mid \mathbf{X}_1 = \mathbf{x}\} \right\}$$
$$= \max_{\mathbf{b}'} \left\{ v(\mathbf{b}, \mathbf{b}', \mathbf{x}) + \mathbb{E}\{F(\mathbf{b}', \mathbf{X}_2)\} \right\}.$$

This problem was studied by [Iyengar (2002, 2005)]. Concerning Theorem 3.1, the conditional expected value was simplified to simple expected value in context of F in (3.14), and their proof permits to remove the last assumption in Condition (ii). Their result is analogous for Theorem 3.2.

Remark 3.4. Use of portfolio $\mathbf{b}_n^*$ in Theorem 3.1 requires a solution of the non-discounted Bellman-equation (3.13). For this, an iteration procedure is given in Lemma 3.2 below.

Remark 3.5. In practice, the conditional expectations are unknown and they can be replaced by estimates. It is an open problem what the loss in growth rate is, if we apply estimates in the Bellman-equation

$$W_c^* + F(\mathbf{b}, \mathbf{x}) = \max_{\mathbf{b}'}\{\log w(\mathbf{b}, \mathbf{b}', \mathbf{x}) + \mathbb{E}\{\ln \langle \mathbf{b}', \mathbf{X}_2 \rangle \mid \mathbf{X}_1 = \mathbf{x}\}$$
$$+ \mathbb{E}\{F(\mathbf{b}', \mathbf{X}_2) \mid \mathbf{X}_1 = \mathbf{x}\}\}.$$

Rule 4. Choose a sequence $0 < \delta_n < 1, n = 1, 2, \ldots$, such that

$$\delta_n \downarrow 0, \quad \sum_n \delta_n = \infty, \quad \frac{\delta_{n+1}}{\delta_n} \to 1 \quad (n \to \infty),$$

e.g., $\delta_n = \frac{1}{n+1}$. Set

$$F_1' \triangleq 0,$$

and iterate

$$F_{n+1}' \triangleq M_{\delta_n} F_n' - \max_{\mathbf{b}, \mathbf{x}}(M_{\delta_n} F_n')(\mathbf{b}, \mathbf{x}) \quad (n = 1, 2, \ldots)$$

with

$$(M_{\delta_n} F)(\mathbf{b}, \mathbf{x}) \triangleq \max_{\tilde{\mathbf{b}}} \left\{ v(\mathbf{b}, \tilde{\mathbf{b}}, \mathbf{x}) + (1 - \delta_n)\mathbb{E}\{F(\tilde{\mathbf{b}}, \mathbf{X}_2) \mid \mathbf{X}_1 = \mathbf{x}\} \right\}, \ F \in C.$$

Put

$$\mathbf{b}_1' = \{1/d, \ldots, 1/d\}$$

and

$$\mathbf{b}_{i+1}' = \arg\max_{\tilde{\mathbf{b}}} \left\{ v(\mathbf{b}_i', \tilde{\mathbf{b}}, \mathbf{X}_i) + (1 - \delta_i)\mathbb{E}\{F_i'(\tilde{\mathbf{b}}, \mathbf{X}_{i+1}) | \mathbf{X}_i\} \right\},$$

for $1 \leq i$. This non-stationary rule can be interpreted as a combination of the value iteration and Rule 1.

Theorem 3.2. *Under the conditions (i), (ii) and (iii), if S'_n denotes the wealth at period n using the portfolio $\{\mathbf{b}'_n\}$ then*

$$\lim_{n \to \infty} \frac{1}{n} \ln S'_n = W_c^* \qquad \text{almost surely.}$$

Note that according to Theorem 3.1, if S_n denotes the wealth at period n using any portfolio $\{\mathbf{b}_n\}$ then

$$\limsup_{n \to \infty} \frac{1}{n} \ln S_n \leq W_c^* \qquad \text{almost surely.}$$

3.5. Portfolio Selection with Consumption

For a real number x, let x^+ be the positive part of x. Assume that, at the end of trading period n, there is a consumption $c_n \geq 0$. For the trading period n, the initial capital is S_{n-1}, therefore,

$$S_n = (S_{n-1} \langle \mathbf{b}_n , \mathbf{x}_n \rangle - c_n)^+ .$$

If $S_j > 0$ for all $j = 1, \ldots, n$ then we show by induction that

$$S_n = S_0 \prod_{i=1}^{n} \langle \mathbf{b}_i , \mathbf{x}_i \rangle - \sum_{k=1}^{n} c_k \prod_{i=k+1}^{n} \langle \mathbf{b}_i , \mathbf{x}_i \rangle , \qquad (3.15)$$

where the empty product is 1, by definition. For $n = 1$, (3.15) holds. Assume (3.15) for $n - 1$:

$$S_{n-1} = S_0 \prod_{i=1}^{n-1} \langle \mathbf{b}_i , \mathbf{x}_i \rangle - \sum_{k=1}^{n-1} c_k \prod_{i=k+1}^{n-1} \langle \mathbf{b}_i , \mathbf{x}_i \rangle .$$

Then,

$$\begin{aligned} S_n &= S_{n-1} \langle \mathbf{b}_n , \mathbf{x}_n \rangle - c_n \\ &= \left(S_0 \prod_{i=1}^{n-1} \langle \mathbf{b}_i , \mathbf{x}_i \rangle - \sum_{k=1}^{n-1} c_k \prod_{i=k+1}^{n-1} \langle \mathbf{b}_i , \mathbf{x}_i \rangle \right) \langle \mathbf{b}_n , \mathbf{x}_n \rangle - c_n \\ &= S_0 \prod_{i=1}^{n} \langle \mathbf{b}_i , \mathbf{x}_i \rangle - \sum_{k=1}^{n} c_k \prod_{i=k+1}^{n} \langle \mathbf{b}_i , \mathbf{x}_i \rangle . \end{aligned}$$

One has to emphasize that (3.15) holds for all n iff $S_n > 0$ for all n, otherwise there is a ruin. In what follows, we study the average growth rate under no ruin and the probability of ruin.

By definition,

$$
\mathbb{P}\{\text{ ruin }\} = \mathbb{P}\left\{\bigcup_{n=1}^{\infty}\{S_n = 0\}\right\}
$$

$$
= \mathbb{P}\left\{\bigcup_{n=1}^{\infty}\left\{S_0\prod_{i=1}^{n}\langle\mathbf{b}_i\,,\,\mathbf{x}_i\rangle - \sum_{k=1}^{n}c_k\prod_{i=k+1}^{n}\langle\mathbf{b}_i\,,\,\mathbf{x}_i\rangle \le 0\right\}\right\},
$$

therefore,

$$
\mathbb{P}\{\text{ ruin }\} = \mathbb{P}\left\{\bigcup_{n=1}^{\infty}\left\{\prod_{i=1}^{n}\langle\mathbf{b}_i\,,\,\mathbf{x}_i\rangle\left(S_0 - \sum_{k=1}^{n}\frac{c_k}{\prod_{i=1}^{k}\langle\mathbf{b}_i\,,\,\mathbf{x}_i\rangle}\right) \le 0\right\}\right\}
$$

$$
\le \mathbb{P}\left\{\bigcup_{n=1}^{\infty}\left\{\prod_{i=1}^{n}\langle\mathbf{b}_i\,,\,\mathbf{x}_i\rangle\left(S_0 - \sum_{k=1}^{\infty}\frac{c_k}{\prod_{i=1}^{k}\langle\mathbf{b}_i\,,\,\mathbf{x}_i\rangle}\right) \le 0\right\}\right\}
$$

$$
\le \mathbb{P}\left\{S_0 \le \sum_{k=1}^{\infty}\frac{c_k}{\prod_{i=1}^{k}\langle\mathbf{b}_i\,,\,\mathbf{x}_i\rangle}\right\} \tag{3.16}
$$

and

$$
\mathbb{P}\{\text{ ruin }\} = \mathbb{P}\left\{\bigcup_{n=1}^{\infty}\left\{\prod_{i=1}^{n}\langle\mathbf{b}_i\,,\,\mathbf{x}_i\rangle\left(S_0 - \sum_{k=1}^{n}\frac{c_k}{\prod_{i=1}^{k}\langle\mathbf{b}_i\,,\,\mathbf{x}_i\rangle}\right) \le 0\right\}\right\}
$$

$$
\ge \max_{n}\mathbb{P}\left\{\prod_{i=1}^{n}\langle\mathbf{b}_i\,,\,\mathbf{x}_i\rangle\left(S_0 - \sum_{k=1}^{n}\frac{c_k}{\prod_{i=1}^{k}\langle\mathbf{b}_i\,,\,\mathbf{x}_i\rangle}\right) \le 0\right\}
$$

$$
= \mathbb{P}\left\{S_0 \le \sum_{k=1}^{\infty}\frac{c_k}{\prod_{i=1}^{k}\langle\mathbf{b}_i\,,\,\mathbf{x}_i\rangle}\right\}. \tag{3.17}
$$

(3.16) and (3.17) imply that

$$
\mathbb{P}\{\text{ ruin }\} = \mathbb{P}\left\{S_0 \le \sum_{k=1}^{\infty}\frac{c_k}{\prod_{i=1}^{k}\langle\mathbf{b}_i\,,\,\mathbf{x}_i\rangle}\right\}.
$$

Under no ruin, on one hand we obtain the upper bound on the average growth rate

$$
\begin{aligned}
W_n &= \frac{1}{n} \ln S_n \\
&= \frac{1}{n} \ln \left(S_0 \prod_{i=1}^{n} \langle \mathbf{b}_i , \mathbf{x}_i \rangle - \sum_{k=1}^{n} c_k \prod_{i=k+1}^{n} \langle \mathbf{b}_i , \mathbf{x}_i \rangle \right) \\
&\leq \frac{1}{n} \ln S_0 \prod_{i=1}^{n} \langle \mathbf{b}_i , \mathbf{x}_i \rangle \\
&= \frac{1}{n} \sum_{i=1}^{n} \ln \langle \mathbf{b}_i , \mathbf{x}_i \rangle + \frac{1}{n} \ln S_0 .
\end{aligned}
$$

On the other hand, we have the lower bound

$$
\begin{aligned}
W_n &= \frac{1}{n} \ln S_n \\
&= \frac{1}{n} \ln \left(S_0 \prod_{i=1}^{n} \langle \mathbf{b}_i , \mathbf{x}_i \rangle - \sum_{k=1}^{n} c_k \prod_{i=k+1}^{n} \langle \mathbf{b}_i , \mathbf{x}_i \rangle \right) \\
&= \frac{1}{n} \ln \prod_{i=1}^{n} \langle \mathbf{b}_i , \mathbf{x}_i \rangle \left(S_0 - \sum_{k=1}^{n} \frac{c_k}{\prod_{i=1}^{k} \langle \mathbf{b}_i , \mathbf{x}_i \rangle} \right) \\
&\geq \frac{1}{n} \ln \prod_{i=1}^{n} \langle \mathbf{b}_i , \mathbf{x}_i \rangle \left(S_0 - \sum_{k=1}^{\infty} \frac{c_k}{\prod_{i=1}^{k} \langle \mathbf{b}_i , \mathbf{x}_i \rangle} \right) \\
&= \frac{1}{n} \sum_{i=1}^{n} \ln \langle \mathbf{b}_i , \mathbf{x}_i \rangle + \frac{1}{n} \ln \left(S_0 - \sum_{k=1}^{\infty} \frac{c_k}{\prod_{i=1}^{k} \langle \mathbf{b}_i , \mathbf{x}_i \rangle} \right) ,
\end{aligned}
$$

therefore, under no ruin, the asymptotic average growth rate with consumption is the same as without consumption:

$$
W_n = \frac{1}{n} \ln S_n \approx \frac{1}{n} \sum_{i=1}^{n} \ln \langle \mathbf{b}_i , \mathbf{x}_i \rangle .
$$

Consider the case of constant consumption, i.e., $c_n = c > 0$. Then, there is no ruin if

$$
S_0 > c \sum_{k=1}^{\infty} \frac{1}{\prod_{i=1}^{k} \langle \mathbf{b}_i , \mathbf{x}_i \rangle} .
$$

Because of the definition of the average growth rate, we have that

$$
W_k \approx \frac{1}{k} \ln \prod_{i=1}^{k} \langle \mathbf{b}_i , \mathbf{x}_i \rangle ,
$$

which implies that

$$\sum_{k=1}^{\infty} \frac{1}{\prod_{i=1}^{k} \langle \mathbf{b}_i, \mathbf{x}_i \rangle} \approx \sum_{k=1}^{\infty} e^{-kW_k}.$$

Assume that our portfolio selection is asymptotically optimal, which means that

$$\lim_{n \to \infty} W_n = W^*.$$

Then,

$$\sum_{k=1}^{\infty} \frac{1}{\prod_{i=1}^{k} \langle \mathbf{b}_i, \mathbf{x}_i \rangle} \approx \sum_{k=1}^{\infty} e^{-kW^*} = \frac{e^{-W^*}}{1 - e^{-W^*}}.$$

This approximation implies that the ruin probability can be small only if

$$S_0 > c \frac{e^{-W^*}}{1 - e^{-W^*}}.$$

A special case of this model is when there is only one risk-free asset

$$S_n = (S_{n-1}(1 + r) - c)^+$$

with some $r > 0$. Obviously, there is no ruin if $S_0 r > c$. It is easy to verify that this assumption can be derived from the general condition if

$$e^{W^*} = 1 + r.$$

The ruin probability can be decreased if the consumptions happen in blocks of size N trading periods. Let S_n denote the wealth at the end of the n-th block. Then,

$$S_n = \left(S_{n-1} \prod_{j=(n-1)N+1}^{nN} \langle \mathbf{b}_j, \mathbf{x}_j \rangle - Nc \right)^+.$$

Similarly to the previous calculations, we can check that under no ruin the average growth rates with and without consumption are the same. Moreover,

$$\mathbb{P}\{ \text{ ruin } \} = \mathbb{P}\left\{ S_0 \leq cN \sum_{k=1}^{\infty} \frac{1}{\prod_{i=1}^{kN} \langle \mathbf{b}_i, \mathbf{x}_i \rangle} \right\}.$$

This ruin probability is a monotonically-decreasing function of N, and, for large N, the exact condition of no ruin is the same as the approximation mentioned above.

This model can be applied to the analysis of portfolio-selection strategies with fixed transaction cost such that c_n is the transaction cost to be paid when changing the portfolio from $\mathbf{b}_n$ to $\mathbf{b}_{n+1}$. In this case, the transaction cost c_n depends on the number of shares involved in the transaction.

Let us calculate c_n. At the end of the n-th trading period and before paying for the transaction cost, the wealth at asset j is $S_{n-1}b_n^{(j)}x_n^{(j)}$, which means that the number of shares j is

$$m_n^{(j)} = \frac{S_{n-1}b_n^{(j)}x_n^{(j)}}{S_n^{(j)}}.$$

In the model of fixed transaction cost, we assume that $m_n^{(j)}$ is integer. If one changes the portfolio $\mathbf{b}_n$ to $\mathbf{b}_{n+1}$, then the wealth at asset j should be $S_{n-1}\langle \mathbf{b}_n , \mathbf{x}_n\rangle b_{n+1}^{(j)}$, so the number of shares j should be

$$m_{n+1}^{(j)} = \frac{S_{n-1}\langle \mathbf{b}_n , \mathbf{x}_n\rangle b_{n+1}^{(j)}}{S_n^{(j)}}.$$

If $m_{n+1}^{(j)} < m_n^{(j)}$ then we have to sell, and the wealth we have is

$$\sum_{j=1}^{d}\left(m_n^{(j)} - m_{n+1}^{(j)}\right)^+ S_n^{(j)} = \sum_{j=1}^{d}\left(S_{n-1}b_n^{(j)}x_n^{(j)} - S_{n-1}\langle \mathbf{b}_n , \mathbf{x}_n\rangle b_{n+1}^{(j)}\right)^+.$$

If $m_{n+1}^{(j)} > m_n^{(j)}$ then we have to buy, and the wealth that we pay is

$$\sum_{j=1}^{d}\left(m_{n+1}^{(j)} - m_n^{(j)}\right)^+ S_n^{(j)} = \sum_{j=1}^{d}\left(S_{n-1}\langle \mathbf{b}_n , \mathbf{x}_n\rangle b_{n+1}^{(j)} - S_{n-1}b_n^{(j)}x_n^{(j)}\right)^+.$$

Let $C > 0$ be the fixed transaction cost. Then, the transaction fee is

$$c_n = c_n(\mathbf{b}_{n+1}) = C\sum_{j=1}^{d}\left|m_n^{(j)} - m_{n+1}^{(j)}\right|.$$

The portfolio selection $\mathbf{b}_{n+1}$ is self-financing if

$$\sum_{j=1}^{d}\left(S_{n-1}b_n^{(j)}x_n^{(j)} - S_{n-1}\langle \mathbf{b}_n , \mathbf{x}_n\rangle b_{n+1}^{(j)}\right)^+$$

$$\geq \sum_{j=1}^{d}\left(S_{n-1}\langle \mathbf{b}_n , \mathbf{x}_n\rangle b_{n+1}^{(j)} - S_{n-1}b_n^{(j)}x_n^{(j)}\right)^+ + c_n.$$

$\mathbf{b}_{n+1}$ is an admissible portfolio if $m_{n+1}^{(j)}$ is integer for all j and it satisfies the self-financing condition. The set of admissible portfolios is denoted by $\Delta_{n,d}$.

Taking into account the fixed transaction cost, a kernel-based portfolio selection can be defined as follows. Choose the radius $r_{k,\ell} > 0$ such that, for any fixed k,

$$\lim_{\ell \to \infty} r_{k,\ell} = 0.$$

For $n > k + 1$, introduce the expert $\mathbf{b}^{(k,\ell)}$ by

$$\mathbf{b}_{n+1}^{(k,\ell)} = \arg\max_{\mathbf{b} \in \Delta_{n,d}} \sum_{i \in J} \ln \left\{ (S_{n-1}^{(k,\ell)} \left\langle \mathbf{b}_n^{(k,\ell)}, \mathbf{x}_n \right\rangle - c_n(\mathbf{b})) \left\langle \mathbf{b}, \mathbf{x}_i \right\rangle \right\},$$

if the sum is non-void, and $\mathbf{b}_0 = (1/d, \ldots, 1/d)$ otherwise, where

$$J = \left\{ k < i \le n : \|\mathbf{x}_{i-k}^{i-1} - \mathbf{x}_{n-k+1}^n\| \le r_{k,\ell} \right\}.$$

Combine the elementary portfolio strategies $\mathbf{B}^{(k,\ell)} = \{\mathbf{b}_n^{(k,\ell)}\}$ as in (3.9).

3.6. Proofs

We split the statement of Theorem 3.1 into two propositions.

Proposition 3.1. *Under the conditions (i), (ii) and (iii) the Bellman-equation (3.13) has a solution (W_c^*, F), such that the function F is bounded and continuous, where*

$$\max_{\mathbf{b},\mathbf{x}} F(\mathbf{b}, \mathbf{x}) = 0.$$

Proof. Let C be the Banach space of continuous functions F defined on the compact set $\Delta_d \times [a_1, a_2]^d$ with the sup norm $\| \cdot \|_\infty$. For $0 \le \delta < 1$ and for $f \in C$, define the operator

$$(M_\delta f)(\mathbf{b}, \mathbf{x}) \triangleq \max_{\mathbf{b}'} \{v(\mathbf{b}, \mathbf{b}', \mathbf{x}) + (1 - \delta)\mathbb{E}\{f(\mathbf{b}', \mathbf{X}_2) \mid \mathbf{X}_1 = \mathbf{x}\}\}.$$

$$(3.18)$$

By continuity assumption (ii), this leads to an operator

$$M_\delta : C \to C.$$

(See [Schäfer (2002)] p.114.)

The operator M_δ is continuous, even Lipschitz-continuous with Lipschitz constant $1 - \delta$. Indeed, for $f, f' \in C$, from the representation

$$(M_\delta f)(\mathbf{b}, \mathbf{x}) = v(\mathbf{b}, \mathbf{b}_f^*(\mathbf{b}, \mathbf{x}), \mathbf{x}) + (1 - \delta)\mathbb{E}\{f(\mathbf{b}_f^*(\mathbf{b}, \mathbf{x}), \mathbf{X}_2) \mid \mathbf{X}_1 = \mathbf{x}\}$$

and from the corresponding representation of $(M_\delta f')(\mathbf{b}, \mathbf{x})$, one obtains

$$
\begin{aligned}
(M_\delta f')(\mathbf{b}, \mathbf{x}) &\geq v(\mathbf{b}, \mathbf{b}_f^*(\mathbf{b}, \mathbf{x}), \mathbf{x}) + (1 - \delta)\mathbb{E}\{f'(\mathbf{b}_f^*(\mathbf{b}, \mathbf{x}), \mathbf{X}_2) \mid \mathbf{X}_1 = \mathbf{x}\} \\
&\geq v(\mathbf{b}, \mathbf{b}_f^*(\mathbf{b}, \mathbf{x}), \mathbf{x}) + (1 - \delta)\mathbb{E}\{f(\mathbf{b}_f^*(\mathbf{b}, \mathbf{x}), \mathbf{X}_2) \mid \mathbf{X}_1 = \mathbf{x}\} \\
&\quad -(1 - \delta)\|f - f'\|_\infty \\
&= (M_\delta f)(\mathbf{b}, \mathbf{x}) - (1 - \delta)\|f - f'\|_\infty
\end{aligned}
$$

for all $(\mathbf{b}, \mathbf{x}) \in \Delta_d \times [a_1, a_2]^d$, therefore,

$$
\|M_\delta f - M_\delta f'\|_\infty \leq (1 - \delta)\|f - f'\|_\infty.
$$

Thus, by Banach's fixed-point theorem, the Bellman-equation

$$
\lambda + F(\mathbf{b}, \mathbf{x}) = \max_{\mathbf{b}'} \{v(\mathbf{b}, \mathbf{b}', \mathbf{x}) + (1 - \delta)\mathbb{E}\{F(\mathbf{b}', \mathbf{X}_2) \mid \mathbf{X}_1 = \mathbf{x}\}\}, \quad (3.19)
$$

i.e.,

$$
\lambda + F = M_\delta F
$$

with $\lambda \in \mathbb{R}$, has a unique solution if $0 < \delta < 1$. (3.19) corresponds to (3.12) for $\lambda = 0$, $0 < \delta < 1$, with the unique solution denoted by F_δ, and to (3.13) for $\lambda = W_c^*$ and $\delta = 0$.

We notice

$$
\sup_{0 < \delta < 1} \delta \|F_\delta\|_\infty \leq \max_{\mathbf{b}, \mathbf{b}', \mathbf{x}} |v(\mathbf{b}, \mathbf{b}', \mathbf{x})| < \infty,
$$

(cf. [Schäfer (2002)], Lemma 4.2.3). Similarly to [Iyengar (2002)], put

$$
m_\delta \triangleq \max_{(\mathbf{b}, \mathbf{x})} F_\delta(\mathbf{b}, \mathbf{x}), \quad (3.20)
$$

where we obtain

$$
\sup_{0 < \delta < 1} \delta m_\delta < \infty.
$$

Put

$$
W_c^* \triangleq \limsup_{\delta \downarrow 0} \delta m_\delta
$$

and

$$
\tilde{F}_\delta(\mathbf{b}, \mathbf{x}) \triangleq F_\delta(\mathbf{b}, \mathbf{x}) - m_\delta. \quad (3.21)
$$

Thus,

$$
\max_{(\mathbf{b}, \mathbf{x})} \tilde{F}_\delta(\mathbf{b}, \mathbf{x}) = 0. \quad (3.22)
$$

Corollary 3.1. *Assume the conditions of Proposition 3.1 and let m_δ be defined by (3.20). Then,*

$$\delta m_\delta \to W_c^* \text{ as } \delta \downarrow 0.$$

For each sequence $0 < \delta_n < 1$ with $\delta_n \downarrow 0$, the sequence $\tilde{F}_{\delta_n} \in C$ defined by (3.21) converges to a set of solutions F of the Bellman-equation (3.13).

Proof. Since, in the proof of Proposition 3.1 $\limsup_{\delta \downarrow 0} \delta m_\delta$ can be replaced by $\liminf_{\delta \downarrow 0} \delta m_\delta$, the uniqueness of W_c^* yields the existence of $\lim_{\delta \downarrow 0} \delta m_\delta = W_c^*$. For each sequence $\delta_n \downarrow 0$, a subsequence δ_{n_ℓ} exists, such that $\tilde{F}_{\delta_{n_\ell}}$ converges in C to some solution F of (3.13). This proves the second assertion. $\square$

For the proof of Theorem 3.2 we need the following lemma:

Lemma 3.2. *Assume Conditions (i), (ii) and (iii). Let δ_n and F_n' be as in Rule 4. Then, F_n' converges in C to a set of solutions F of the Bellman-equation (3.13). Further,*

$$w_n \triangleq \max_{\mathbf{b}, \mathbf{x}} (M_{\delta_n} F_n')(\mathbf{b}, \mathbf{x}) \to W_c^* \text{ as } n \to \infty.$$

Proof. We can write

$$F_{n+1}' = M_{\delta_n} F_n' - w_n \tag{3.32}$$

with the continuous operator $M_{\delta_n} : C \to C$ according to (3.18). It holds that

$$\begin{aligned}
|F_{n+1}'(\bar{\mathbf{b}}, \bar{\mathbf{x}}) - F_{n+1}'(\mathbf{b}, \mathbf{x})| &= |(M_{\delta_n} F_n')(\bar{\mathbf{b}}, \bar{\mathbf{x}}) - (M_{\delta_n} F_n')(\mathbf{b}, \mathbf{x})| \\
&\leq |(M_{\delta_n} F_n')(\bar{\mathbf{b}}, \bar{\mathbf{x}}) - (M_{\delta_n} F_n')(\mathbf{b}, \bar{\mathbf{x}})| \\
&\quad + |(M_{\delta_n} F_n')(\mathbf{b}, \bar{\mathbf{x}}) - (M_{\delta_n} F_n')(\mathbf{b}, \mathbf{x})| \\
&\leq \max_{\mathbf{b}'} |v(\bar{\mathbf{b}}, \mathbf{b}', \bar{\mathbf{x}}) - v(\mathbf{b}, \mathbf{b}', \bar{\mathbf{x}})| \\
&\quad + \max_{\mathbf{b}'} |v(\mathbf{b}, \mathbf{b}', \bar{\mathbf{x}}) - v(\mathbf{b}, \mathbf{b}', \mathbf{x})| \\
&\quad + \max_{\mathbf{x}, \bar{\mathbf{x}}} V(\mathbf{x}, \bar{\mathbf{x}}) \|F_n'\|_\infty, \tag{3.33}
\end{aligned}$$

where the inequalities are obtained as in the proof of Lemma 3.1. Noticing

$$\max_{\mathbf{b}, \mathbf{x}} F_n'(\mathbf{b}, \mathbf{x}) = 0$$

and, thus,

$$\max_{(\mathbf{b},\mathbf{x}),(\bar{\mathbf{b}},\bar{\mathbf{x}})} |F'_n(\mathbf{b},\mathbf{x}) - F'_n(\bar{\mathbf{b}},\bar{\mathbf{x}})| = \|F'_n\|_\infty.$$

Moreover, the boundedness of v implies that

$$\|F'_{n+1}\|_\infty \le const + \max_{\mathbf{x},\bar{\mathbf{x}}} V(\mathbf{x},\bar{\mathbf{x}}) \|F'_n\|_\infty$$

with $const < \infty$. Then, by induction,

$$\|F'_n\|_\infty \le \frac{const}{1 - \max_{\mathbf{x},\bar{\mathbf{x}}} V(\mathbf{x},\bar{\mathbf{x}})} \triangleq K < \infty. \tag{3.34}$$

It can be easily checked that

$$\|M_{\delta_{n+1}} F'_{n+1} - M_{\delta_n} F'_{n+1}\|_\infty \le (\delta_n - \delta_{n+1}) \|F'_{n+1}\|_\infty. \tag{3.35}$$

According to the proof of Proposition 3.1, the operator M_{δ_n} is Lipschitz-continuous with Lipschitz constant $1 - \delta_n$. Then,

$$\|F'_{n+2} - F'_{n+1}\|_\infty$$
$$= \|M_{\delta_{n+1}} F'_{n+1} - M_{\delta_n} F'_n\|_\infty$$
$$\le \|M_{\delta_n} F'_{n+1} - M_{\delta_n} F'_n\|_\infty + \|M_{\delta_{n+1}} F'_{n+1} - M_{\delta_n} F'_{n+1}\|_\infty$$
$$\le (1 - \delta_n) \|F'_{n+1} - F'_n\|_\infty + \left(1 - \frac{\delta_{n+1}}{\delta_n}\right) \delta_n K.$$

By the condition on δ_n, we then obtain

$$\|F'_{n+1} - F'_n\|_\infty \to 0 \text{ as } n \to \infty, \tag{3.36}$$

(cf. Lemma 1(c) in [Walk and Zsidó (1989)]). Now, let (δ_{n_k}) be an arbitrary subsequence of (δ_n). From (3.33) and (3.34) and Condition (ii), we obtain

$$\sup_i |F'_i(\bar{\mathbf{b}},\bar{\mathbf{x}}) - F'_i(\mathbf{b},\mathbf{x})| \to 0$$

when $(\bar{\mathbf{b}},\bar{\mathbf{x}}) \to (\mathbf{b},\mathbf{x})$, even uniformly with respect to $(\mathbf{b},\mathbf{x})$. This, together with (3.34), yields the existence of a subsequence $(\delta_{n_{k_\ell}})$ and of a function $\bar{F} \in C$ (bounded, where $\max_{\mathbf{b},\mathbf{x}} \bar{F}(\mathbf{b},\mathbf{x}) = 0$) such that

$$\|F'_{n_{k_\ell}} - \bar{F}\|_\infty \to 0 \text{ as } \ell \to \infty. \tag{3.37}$$

Thus, by continuity of M_0,

$$\|M_0 F'_{n_{k_\ell}} - M_0 \bar{F}\|_\infty \to 0 \text{ as } \ell \to \infty. \tag{3.38}$$

By (3.32),

$$F'_{n_{k_\ell}} + (F'_{n_{k_\ell}+1} - F'_{n_{k_\ell}}) = M_0 F'_{n_{k_\ell}} + (M_{\delta_{n_{k_\ell}}} F'_{n_{k_\ell}} - M_0 F'_{n_{k_\ell}}) - w_{n_{k_\ell}}.$$

(3.36) implies that

$$\|F'_{n_{k_\ell}+1} - F'_{n_{k_\ell}}\|_\infty \to 0.$$

By (3.24) and (3.34),

$$\|M_{\delta_{n_{k_\ell}}} F'_{n_{k_\ell}} - M_0 F'_{n_{k_\ell}}\|_\infty \leq \delta_{n_{k_\ell}} K \to 0.$$

This, together with (3.37) and (3.38) yields the convergence of $(w_{n_{k_\ell}})$ and

$$\lim_\ell w_{n_{k_\ell}} + \bar{F} = M_0 \bar{F}.$$

This means that $\bar{F}$ solves the Bellman-equation (3.13) such that $\lim_\ell w_{n_{k_\ell}} = W_c^*$ (unique by Proposition 3.2). These convergence results yield the assertion. $\qquad \square$

Proof of Theorem 3.2. According to Proposition 3.2 and its proof, it is enough to show

$$\lim_{n \to \infty} \frac{1}{n} \sum_{i=1}^n v(\mathbf{b}'_i, \mathbf{b}'_{i+1}, \mathbf{X}_i) = W_c^* \qquad \text{almost surely.} \qquad (3.39)$$

Rule 4 yields

$$
\begin{aligned}
& w_n + F'_{n+1}(\mathbf{b}'_n, \mathbf{X}_n) \\
& = v(\mathbf{b}'_n, \mathbf{b}'_{n+1}, \mathbf{X}_n) + (1 - \delta_n)\mathbb{E}\{F'_n(\mathbf{b}'_{n+1}, \mathbf{X}_{n+1}) \mid \mathbf{b}'_{n+1}, \mathbf{X}_n\},
\end{aligned}
$$

where

$$w_n = \max_{\mathbf{b},\mathbf{x}}(M_{\delta_n} F'_n)(\mathbf{b}, \mathbf{x}).$$

Then,

$$
\frac{1}{n} \sum_{i=1}^{n} v(\mathbf{b}_i', \mathbf{b}_{i+1}', \mathbf{X}_i) = \frac{1}{n} \sum_{i=1}^{n} w_i + \frac{1}{n} \sum_{i=1}^{n} \left(F_{i+1}'(\mathbf{b}_i', \mathbf{X}_i) \right.
$$
$$
\left. -(1 - \delta_i) \mathbb{E}\{ F_i'(\mathbf{b}_{i+1}', \mathbf{X}_{i+1}) \mid \mathbf{b}_{i+1}', \mathbf{X}_i \} \right)
$$
$$
= \frac{1}{n} \sum_{i=1}^{n} w_i
$$
$$
+ \frac{1}{n} \sum_{i=1}^{n} \left(F_i'(\mathbf{b}_{i+1}', \mathbf{X}_{i+1}) - \mathbb{E}\{ F_i'(\mathbf{b}_{i+1}', \mathbf{X}_{i+1}) \mid \mathbf{X}_1^i \} \right)
$$
$$
+ \left[\frac{1}{n} \sum_{i=1}^{n} \left(F_{i+1}'(\mathbf{b}_i', \mathbf{X}_i) - F_i'(\mathbf{b}_{i+1}', \mathbf{X}_{i+1}) \right) \right.
$$
$$
\left. + \frac{1}{n} \sum_{i=1}^{n} \delta_i \mathbb{E}\{ F_i'(\mathbf{b}_{i+1}', \mathbf{X}_{i+1}) \mid \mathbf{X}_1^i \} \right]
$$
$$
\triangleq A_n + B_n + C_n.
$$

By Lemma 3.2, $A_n \to W_c^*$. By (3.34) and Chow's theorem, $B_n \to 0$ almost surely. Further,

$$
|C_n| \le \frac{1}{n} \left| \sum_{i=1}^{n-1} \left(F_{i+2}'(\mathbf{b}_{i+1}', \mathbf{X}_{i+1}) - F_i'(\mathbf{b}_{i+1}', \mathbf{X}_{i+1}) \right) \right|
$$
$$
+ \frac{1}{n} |F_2'(\mathbf{b}_1', \mathbf{X}_1)| + \frac{1}{n} |F_n'(\mathbf{b}_{n+1}', \mathbf{X}_{n+1})| + \frac{1}{n} \sum_{i=1}^{n} \delta_i K
$$
$$
\to 0
$$

by (3.34) and (3.36) and $\delta_n \to 0$. Thus, (3.39) is obtained. $\qquad \square$

References

Akien, M., Sulem, A. and Taksar, M. I. (2001). Dynamic optimization of long-term growth rate for a portfolio with transaction costs and logaritmic utility, *Mathematical Finance* **11**, pp. 153–188.

Algoet, P. and Cover, T. (1988). Asymptotic optimality and asymptotic equipartition properties of log-optimum investments, *Annals of Probability* **16**, pp. 876–898.

Bellman, R. (1957). *Dynamic Programming* (Princeton University Press, Princeton).

Bobryk, R. V. and Stettner, L. (1999). Discrete time portfolio selection with proportional transaction costs. *Probability and Mathematical Statistics* **19**, pp. 235–248.

Davis, M. H. A. and Norman, A. R. (1990). Portfolio selection with transaction costs, *Mathematics of Operations Research* **15**, pp. 676–713.

Eastham, J. and Hastings, K. (1988). Optimal impulse control of portfolios, *Mathematics of Operations Research* **13**, pp. 588–605.

Fernholz, E. R. (2000). *Stochastic Portfolio Theory* (Springer, New York).

Györfi, L., Lugosi, G. and Udina, F. (2006). Nonparametric kernel based sequential investment strategies, *Mathematical Finance* **16**, pp. 337–357.

Györfi, L. and Schäfer, D. (2003). Nonparametric prediction, in J. A. K. Suykens, G. Horváth, S. Basu, C. Micchelli and J. Vandevalle (eds.), *Advances in Learning Theory: Methods, Models and Applications* (IOS Press, NATO Science Series), pp. 339–354.

Györfi, L., Udina, F. and Walk, H. (2008). Nonparametric nearest-neighbor-based empirical portfolio selection strategies, *Statistics and Decisions* **22**, pp. 145–157.

Györfi, L. and Vajda, I. (2008). On the growth of sequential portfolio selection with transaction cost, in Y. Freund, L. Györfi, G. Turn and T. Zeugmann (eds.), *Algorithmic Learning Theory* (Springer-Verlag, Berlin), pp. 108–122.

Hardy, G. H. (1949). *Divergent Series* (Oxford University Press, London).

Hernández-Lerma, O. and Lasserre, J. B. (1996). *Discrete-Time Markov Control Processes: Basic Optimality Criteria* (Springer, New York).

Hinderer, K. (1970). *Foundations of Non-Stationary Dynamic Programming with Discrete Time Parameter* (Springer-Verlag, Berlin).

Iyengar, G. (2002). Discrete time growth optimal investment with costs, URL http://www.columbia.edu/~gi10/Papers/stochastic.pdf.

Iyengar, G. (2005). Universal investment in markets with transaction costs, *Mathematical Finance* **15**, pp. 359–371.

Iyengar, G. and Cover, T. (2000). Growth optimal investment in horse race markets with costs. *IEEE Transactions on Information Theory* **46**, pp. 2675–2683.

Kalai, A. and Blum, A. (1999). Universal portfolios with and without transaction costs, *Machine Learning* **35**, 3, pp. 193–205.

Korn, R. (1998). Portfolio optimization with strictly positive transaction cost and impulse control, *Finance and Stochastics* **2**, pp. 85–114.

Merhav, N., Ordentlich, E., Seroussi, G. and Weinberger, M. J. (2002). On sequential strategies for loss functions with memory, *IEEE Transactions on Information Theory* **48**, pp. 1947–1958.

Morton, A. J. and Pliska, S. R. (1995). Optimal portfolio manangement with transaction costs, *Mathematical Finance* **5**, pp. 337–356.

Palczewski, J. and Stettner, L. (2006). Maximization of portfolio growth rate under fixed and proportional transaction cost, *Communications in Information and Systems* **7**, pp. 31–58.

Pliska, S. R. and Suzuki, K. (2004). Optimal tracking for asset allocation with fixed and proportional transaction costs, *Quantitative Finance* **4**, pp. 223–243.

Sass, J. and Schäl, M. (2010). The numéraire portfolio under proportional transaction cost, Working paper.

Schäfer, D. (2002). *Nonparametric Estimation for Financial Investment under Log-Utility*, Ph.D. thesis, Universität Stuttgart, Shaker Verlag.

Shreve, S. E., Soner, H. and Xu, G. (1991). Optimal investment and consumption with two bonds and transaction costs, *Mathematical Finance* **1**, pp. 53–84.

Shreve, S. E. and Soner, H. M. (1994). Optimal investment and consumption with transaction costs, *Annals of Applied Probability* **4**, pp. 609–692.

Stout, W. F. (1974). *Almost sure convergence* (Academic Press, New York).

Taksar, M., Klass, M. and Assaf, D. (1988). A diffusion model for optimal portfolio selection in the presence of brokerage fees, *Mathematics of Operations Research* **13**, pp. 277–294.

Walk, H. and Zsidó, L. (1989). Convergence of the Robbins–Monro method for linear problems in a Banach space, *Journal of Mathematical Analysis and Applications* **139**, pp. 152–177.

Yosida, K. (1968). *Functional Analysis*, 2nd edn. (Springer-Verlag, Berlin).

Chapter 4

Growth-Optimal Portfolio Selection with Short Selling and Leverage

Márk Horváth* and András Urbán†

Department of Computer Science and Information Theory,
Budapest University of Technology and Economics,
H-1117, Magyar tudósok körútja 2., Budapest, Hungary.
*mhorvath@math.bme.hu, †urbi@cs.bme.hu

The growth-optimal strategy on non-leveraged, long-only memoryless markets is the best constantly-rebalanced portfolio (BCRP), also called the log-optimal strategy. Optimality conditions are derived for frameworks on leverage and short selling, and generalizing the BCRP by establishing no-ruin conditions. Moreover, the strategy and its asymptotic growth rate are investigated under a memoryless assumption, both from theoretical and empirical points of view. The empirical performance of the methods was tested for NYSE data, demonstrating spectacular gains for leveraged portfolios and showing the unimportance of short selling in the growth-rate sense, both in the case of BCRP and dynamic portfolios.

4.1. Introduction

Earlier results in the nonparametric statistics, information theory and economics literature (such as [Kelly (1956)], [Latané (1959)], [Breiman (1961)], [Markowitz (1952)], [Markowitz (1976)] and [Finkelstein and Whitley (1981)]) established the optimality criterion for long-only, non-leveraged investment. These results have shown that the market is inefficient, i.e., substantial gain is achievable by rebalancing and predicting market returns based on the market's history. Our aim is to show that using leverage through margin buying (the act of borrowing money and increasing market exposure) yields substantially higher growth rates in the case of a memoryless (independent identically distributed, i.i.d.) assumption on returns. Besides a framework for leveraged investment, we also establish a mathematical basis for short selling, i.e., creating negative exposure to asset prices. Short selling means the process of borrowing assets and selling them immediately, with the obligation to rebuy them later.

It can be shown that the optimal asymptotic growth rate for a memoryless market coincides with that of the best constantly-rebalanced portfolio (BCRP). The idea is that, on a frictionless market, the investor can rebalance his portfolio for free in each trading period. Hence, asymptotic optimization on a memoryless market means that the growth-optimal strategy will pick the same portfolio vector in each trading period. Strategies based on this observation are called constantly-rebalanced portfolios (CRP), while the one with the highest asymptotic average growth rate is referred to as BCRP. Our results include the generalization of BCRP for margin-buying and short-selling frameworks.

To allow short selling and leverage, our formulation weakens the constraints on the feasible set of possible portfolio vectors, thus, they are expected to improve performance. Leverage is anticipated to have substantial merit in terms of growth rate, while short selling is not expected to yield much better results. We do not expect increased profits on a short-selling allowed CRP strategy during the experiments, since companies worth short selling in our test period had already defaulted by now and hence they are not in our data set. Nonetheless, short selling might yield increased profits in the case of markets with memory, since earlier results have shown that the market was inefficient (cf. [Györfi *et al.* (2006)]). In case of i.i.d. returns with known distribution, [Cover (1984)] has introduced a gradient-based method for optimization of long-only, log-optimal portfolios, and gave necessary and sufficient conditions on growth-optimal investment in [Bell and Cover (1980)]. We extend these results to short selling and leverage.

Contrary to non-leveraged, long-only investment in earlier literature, in the case of margin buying and short selling, it is easy to default on the total initial investment. In this case, asymptotic growth rate is minus infinity. By bounding possible market returns, we establish circumstances such that default is impossible. We do this in such a way that debt and positions are limited and the investor is always able to satisfy his liabilities by selling assets. Restriction of market exposure and amount of debt is in line with the practice of brokerages and regulators.

Our notation for asset prices and returns is as follows. Consider a market consisting of d assets. Evolution of prices is represented by a sequence of price vectors $\mathbf{s}_1, \mathbf{s}_2, \ldots \in \mathbb{R}^d_+$, where

$$\mathbf{s}_n = (s_n^{(1)}, \ldots, s_n^{(d)}). \tag{4.1}$$

$s_n^{(j)}$ denotes the price of the j-th asset at the end of the n-th trading period. In order to apply the usual techniques for time series analysis, we transform the sequence of price vectors $\{\mathbf{s}_n\}$ into return vectors:

$$\mathbf{x}_n = (x_n^{(1)}, \ldots, x_n^{(d)}),$$

where

$$x_n^{(j)} = \frac{s_n^{(j)}}{s_{n-1}^{(j)}}.$$

Here, the j-th component $x_n^{(j)}$ of the return vector $\mathbf{x}_n$ denotes the amount obtained by investing unit capital in the j-th asset during the n-th trading period.

4.2. Non-Leveraged, Long-Only Investment

A representative example of the dynamic portfolio selection in the long-only case is the constantly-rebalanced portfolio (CRP), introduced and studied by [Kelly (1956)], [Latané (1959)], [Breiman (1961)], [Markowitz (1976)], [Finkelstein and Whitley (1981)], [Móri (1982)], [Móri and Székely (1982)] and [Cover (1984)]. For a comprehensive survey, see also Chapters 6 and 15 in [Cover and Thomas (1991)], and Chapter 15 in [Luenberger (1998)].

CRP is a self-financing portfolio strategy, rebalancing to the same proportional portfolio in each investment period. This means that the investor neither consumes from, nor deposits new cash into his account, but reinvests his capital in each trading period. Using this strategy, the investor chooses a proportional portfolio vector $\mathbf{b} = (b^{(1)}, \ldots, b^{(d)})$, and rebalances his portfolio after each period to correct the price shifts in the market. This way, the proportion of his wealth invested in each asset at the beginning of trading periods is constant.

The j-th component $b^{(j)}$ of $\mathbf{b}$ denotes the proportion of the investor's capital invested in asset j. Thus the portfolio vector has nonnegative components that sum up to 1. The set of portfolio vectors is denoted by

$$\Delta_d = \left\{ \mathbf{b} = (b^{(1)}, \ldots, b^{(d)});\ b^{(j)} \geq 0,\ \sum_{j=1}^{d} b^{(j)} = 1 \right\}. \tag{4.2}$$

Let S_0 denote the investor's initial capital. At the beginning of the first trading period $S_0 b^{(j)}$ is invested into asset j, and it results in position size

$S_0 b^{(j)} x_1^{(j)}$ after changes in market prices. Therefore, at the end of the first trading period the investor's wealth becomes

$$S_1 = S_0 \sum_{j=1}^{d} b^{(j)} x_1^{(j)} = S_0 \langle \mathbf{b} , \mathbf{x}_1 \rangle ,$$

where $\langle \cdot , \cdot \rangle$ denotes the inner product. For the second trading period, S_1 is the new initial capital, hence

$$S_2 = S_1 \langle \mathbf{b} , \mathbf{x}_2 \rangle = S_0 \langle \mathbf{b} , \mathbf{x}_1 \rangle \langle \mathbf{b} , \mathbf{x}_2 \rangle .$$

By induction, for trading period n,

$$S_n = S_{n-1} \langle \mathbf{b} , \mathbf{x}_n \rangle = S_0 \prod_{i=1}^{n} \langle \mathbf{b} , \mathbf{x}_i \rangle .$$

Including a cash account into the framework is straightforward by assuming

$$x_n^{(j)} = 1$$

for some j and for all n. The asymptotic average growth rate of this portfolio selection is

$$\begin{aligned}
W(\mathbf{b}) &= \lim_{n \to \infty} \ln \sqrt[n]{S_n} = \lim_{n \to \infty} \frac{1}{n} \ln S_n \\
&= \lim_{n \to \infty} \left(\frac{1}{n} \ln S_0 + \frac{1}{n} \sum_{i=1}^{n} \ln \langle \mathbf{b} , \mathbf{x}_i \rangle \right) \\
&= \lim_{n \to \infty} \frac{1}{n} \sum_{i=1}^{n} \ln \langle \mathbf{b} , \mathbf{x}_i \rangle ,
\end{aligned}$$

if the limit exists. This also means that, without loss of generality, we can assume that the initial capital $S_0 = 1$.

If the market process $\{\mathbf{X}_i\}$ is memoryless, i.e., is a sequence of i.i.d. random return vectors, then the asymptotic rate of growth exists almost surely, where, with random vector X being distributed as X_i,

$$W(\mathbf{b}) = \lim_{n \to \infty} \frac{1}{n} \sum_{i=1}^{n} \ln \langle \mathbf{b} , \mathbf{X}_i \rangle = \mathbb{E} \ln \langle \mathbf{b} , \mathbf{X} \rangle \qquad \text{almost surely,} \qquad (4.3)$$

given that $\mathbb{E} \ln \langle \mathbf{b} , \mathbf{X} \rangle$ is finite, due to strong law of large numbers. We can ensure this property by assuming finiteness of $\mathbb{E} \ln X^{(j)}$, i.e., $\mathbb{E}| \ln X^{(j)}| < \infty$ for each $j \in \{1, \ldots, d\}$.

In fact, because of $b^{(i)} > 0$ for some i, we have

$$\mathbb{E}\ln\langle \mathbf{b}, \mathbf{X}\rangle \geq \mathbb{E}\ln\left(b^{(i)} X^{(j)}\right)$$
$$= \ln b^{(i)} + \mathbb{E}\ln X^{(i)} > -\infty,$$

and, because of $b^{(j)} \leq 1$ for all j, we have

$$\mathbb{E}\ln\langle \mathbf{b}, \mathbf{X}\rangle \leq \mathbb{E}\ln\left(d\max_j X^{(j)}\right)$$
$$= \ln d + \mathbb{E}\max_j \ln X^{(j)}$$
$$\leq \ln d + \mathbb{E}\max_j \ln |X^{(j)}|$$
$$\leq \ln d + \sum_j \mathbb{E}\ln|X^{(j)}| < \infty.$$

From (4.3) it follows that rebalancing according to the best log-optimal strategy

$$\mathbf{b}^* \in \arg\max_{\mathbf{b}\in\Delta_d} \mathbb{E}\ln\langle \mathbf{b}, \mathbf{X}\rangle,$$

is also an asymptotically-optimal trading strategy, i.e., a strategy with optimum asymptotic growth

$$W(\mathbf{b}^*) \geq W(\mathbf{b}) \qquad \text{almost surely,}$$

for any $\mathbf{b} \in \Delta_d$. The strategy of rebalancing according to $\mathbf{b}^*$ at the beginning of each trading period is called the best constantly-rebalanced portfolio (BCRP).

In the following, we repeat calculations of [Bell and Cover (1980)]. Our aim is to maximize the asymptotic average rate of growth. $W(\mathbf{b})$ being concave, we minimize the convex objective function

$$f_{\mathbf{X}}(\mathbf{b}) = -W(\mathbf{b}) = -\mathbb{E}\ln\langle \mathbf{b}, \mathbf{X}\rangle.$$

To use the Kuhn–Tucker theorem, we establish linear, inequality-type constraints over the search space Δ_d in (4.2):

$$-b^{(i)} \leq 0,$$

for $i = 1, \ldots, d$, i.e.

$$\langle \mathbf{b}, \mathbf{a}_i\rangle \leq 0,$$

where $\mathbf{a}_i \in \mathbb{R}^d$ denotes the i-th unit vector, having -1 at position i.

Our only equality-type constraint is

$$\sum_{j=1}^{d} b^{(j)} - 1 = 0,$$

i.e.

$$\langle \mathbf{b}, \mathbf{e} \rangle - 1 = 0, \tag{4.4}$$

where $\mathbf{e} \in \mathbb{R}^d, \mathbf{e} = (1, 1, \ldots, 1)$.

The partial derivatives of the objective function are

$$\frac{\partial f_{\mathbf{X}}(\mathbf{b})}{\partial b^{(i)}} = -\mathbb{E}\frac{X^{(i)}}{\langle \mathbf{b}, \mathbf{X} \rangle},$$

for $i = 1, \ldots, d$.

According to the Kuhn–Tucker theorem ([Kuhn and Tucker (1951)]), the portfolio vector $\mathbf{b}^*$ is optimal, if and only if, there are constants $\mu_i \geq 0$ $(i = 1, \ldots, d)$ and $\vartheta \in \mathbb{R}$, such that

$$f_{\mathbf{X}}'(\mathbf{b}^*) + \sum_{i=1}^{d} \mu_i \mathbf{a}_i + \vartheta \mathbf{e} = 0$$

and

$$\mu_j \langle \mathbf{b}^*, \mathbf{a}_j \rangle = 0,$$

for $j = 1, \ldots, d$.

This means that

$$-\mathbb{E}\frac{X^{(j)}}{\langle \mathbf{b}^*, \mathbf{X} \rangle} - \mu_j + \vartheta = 0 \tag{4.5}$$

and

$$\mu_j b^{*(j)} = 0,$$

for $j = 1, \ldots, d$. Summing up (4.5) weighted by $b^{*(j)}$, we obtain:

$$-\mathbb{E}\frac{\langle \mathbf{b}^*, \mathbf{X} \rangle}{\langle \mathbf{b}^*, \mathbf{X} \rangle} - \sum_{j=1}^{d} \mu_j b^{*(j)} + \sum_{j=1}^{d} \vartheta b^{*(j)} = 0,$$

hence,

$$\vartheta = 1.$$

We can state the following necessary condition for optimality of $\mathbf{b}^*$. If

$$\mathbf{b}^* \in \arg \max_{\mathbf{b} \in \Delta_d} W(\mathbf{b}),$$

then

$$b^{*(j)} > 0 \implies \mu_j = 0 \implies \mathbb{E}\frac{X^{(j)}}{\langle \mathbf{b}^*, \mathbf{X} \rangle} = 1, \qquad (4.6)$$

and

$$b^{*(j)} = 0 \implies \mu_j \geq 0 \implies \mathbb{E}\frac{X^{(j)}}{\langle \mathbf{b}^*, \mathbf{X} \rangle} \leq 1. \qquad (4.7)$$

Because of the convexity of $f_\mathbf{X}(\mathbf{b})$, the former conditions are also sufficient. Assume $\mathbf{b}^* \in \Delta_d$. If, for any fixed $j = 1, \ldots, d$, either

$$\mathbb{E}\frac{X^{(j)}}{\langle \mathbf{b}^*, \mathbf{X} \rangle} = 1 \text{ and } b^{*(j)} > 0,$$

or

$$\mathbb{E}\frac{X^{(j)}}{\langle \mathbf{b}^*, \mathbf{X} \rangle} \leq 1 \text{ and } b^{*(j)} = 0,$$

then $\mathbf{b}^*$ is optimal. The latter two conditions pose a necessary and sufficient condition on optimality of b^*.

Remark 4.1. In the case of an *independent* asset, i.e., for some $j \in 1, \ldots, d$, $X^{(j)}$ being independent from the rest of the assets,

$$b^{*(j)} = 0 \implies \mathbb{E}\frac{X^{(j)}}{\langle \mathbf{b}^*, \mathbf{X} \rangle} \leq 1$$

implies by $b^{*(j)} = 0$ that $X^{(j)}$ is independent of $\langle \mathbf{b}^*, \mathbf{X} \rangle$. This means that

$$b^{*(j)} = 0 \implies \mathbb{E}X^{(j)}\mathbb{E}\frac{1}{\langle \mathbf{b}^*, \mathbf{X} \rangle} \leq 1,$$

therefore,

$$b^{*(j)} = 0 \implies \mathbb{E}X^{(j)} \leq \frac{1}{\mathbb{E}\frac{1}{\langle \mathbf{b}^*, \mathbf{X} \rangle}}.$$

According to the Kuhn–Tucker theorem, for any fixed $j = 1, \ldots, d$, either

$$\mathbb{E}\frac{X^{(j)}}{\langle \mathbf{b}^*, \mathbf{X} \rangle} = 1 \text{ and } b^{*(j)} > 0,$$

or

$$\mathbb{E}X^{(j)} \leq \frac{1}{\mathbb{E}\frac{1}{\langle \mathbf{b}^*, \mathbf{X} \rangle}} \text{ and } b^{*(j)} = 0,$$

if and only if $\mathbf{b}^*$ is optimal, i.e.,

$$\mathbf{b}^* \in \arg \max_{\mathbf{b} \in \Delta_d} W(\mathbf{b}). \qquad (4.8)$$

Remark 4.2. Assume the optimal portfolio $\mathbf{b}^*$ for d assets

$$\mathbf{X} = (X^{(1)}, X^{(2)}, \ldots, X^{(d)})$$

is already established. Given a new asset, independent of our previous d assets, we can formulate a condition on its inclusion in the new optimal portfolio $\mathbf{b}^{**}$. If

$$\mathbb{E}X^{(d+1)} < \frac{1}{\mathbb{E}\frac{1}{\langle \mathbf{b}^*, \mathbf{X}\rangle}}$$

then

$$b^{**(d+1)} = 0.$$

This means that, for a new independent asset like cash, we do not have to do the *optimization* for each asset in the portfolio, and we can achieve a substantial reduction in the dimension of the search for an optimal portfolio.

Remark 4.3. The same trick can be applied in case of *dependent* returns as well. If

$$\mathbb{E}\frac{X^{(d+1)}}{\langle \mathbf{b}^*, \mathbf{X}\rangle} < 1 \text{ then } b^{**(d+1)} = 0,$$

which is much simpler to verify then performing *optimization* of asymptotic average growth. This condition can be formulated as

$$\mathbb{E}X^{(d+1)}\mathbb{E}\frac{1}{\langle \mathbf{b}^*, \mathbf{X}\rangle} + \mathbf{Cov}\left(X^{(d+1)}, \frac{1}{\langle \mathbf{b}^*, \mathbf{X}\rangle}\right) \leq 1,$$

which poses a condition on covariance and expected value of the new asset.

Remark 4.4. [Roll (1973)], [Pulley (1983)] and [Vajda (2006)] suggested an approximation of $\mathbf{b}^*$ using

$$\ln z \approx h(z) = z - 1 - \frac{1}{2}(z - 1)^2,$$

which is the second-order Taylor approximation of the function $\ln z$ at $z = 1$. Then, the *semi-log-optimal* portfolio selection is

$$\overline{\mathbf{b}} \in \arg\max_{\mathbf{b}\in\Delta_d} \mathbb{E}\{h\langle \mathbf{b}, \mathbf{X}\rangle\}.$$

Our new objective function is convex:

$$\begin{aligned}
\overline{f}_{\mathbf{X}}(\mathbf{b}) &= -\mathbb{E}\{h\langle\mathbf{b},\mathbf{X}\rangle\} \\
&= -\mathbb{E}\{\langle\mathbf{b},\mathbf{X}\rangle - 1 - \frac{1}{2}(\langle\mathbf{b},\mathbf{X}\rangle - 1)^2\} \\
&= \mathbb{E}\{-\langle\mathbf{b},\mathbf{X}\rangle + 1 + \frac{1}{2}(\langle\mathbf{b},\mathbf{X}\rangle - 1)^2\} \\
&= \mathbb{E}\{-\langle\mathbf{b},\mathbf{X}\rangle + 1 + \frac{1}{2}\langle\mathbf{b},\mathbf{X}\rangle^2 - \langle\mathbf{b},\mathbf{X}\rangle + \frac{1}{2}\} \\
&= \mathbb{E}\{\frac{1}{2}\langle\mathbf{b},\mathbf{X}\rangle^2 - 2\langle\mathbf{b},\mathbf{X}\rangle + \frac{3}{2}\} \\
&= \frac{1}{2}\langle\mathbf{b},\mathbb{E}(\mathbf{X}\mathbf{X}^T)\mathbf{b}\rangle - \langle\mathbf{b},2\mathbb{E}\mathbf{X}\rangle + \frac{3}{2} \\
&= \mathbb{E}\left\{\left(\frac{1}{\sqrt{2}}\langle\mathbf{b},\mathbf{X}\rangle - \sqrt{2}\right)^2 - \frac{1}{2}\right\},
\end{aligned}$$

where $\mathbf{X}$ is the column vector, $\mathbf{X}\mathbf{X}^T$ denotes the outer product. This is equivalent to minimizing

$$\overline{\overline{f}}_{\mathbf{X}}(\mathbf{b}) = \mathbb{E}(\langle\mathbf{b},\mathbf{X}\rangle - 2)^2.$$

Thus, $\overline{\mathbf{b}}$ can be simply calculated as the minimizer of the squared distance from 2. In the case of data-driven algorithms, the solution is using linear regression, under the constraint $\mathbf{b}^* \in \Delta_d$:

$$\begin{aligned}
\overline{\overline{f}}_{\mathbf{X}}(\mathbf{b}) &= \mathbf{Var}\langle\mathbf{b},\mathbf{X}\rangle + (2 - \mathbb{E}\langle\mathbf{b},\mathbf{X}\rangle)^2 \\
&= \mathbf{Var}\langle\mathbf{b},\mathbf{X}\rangle + (2 - \langle\mathbf{b},\mathbb{E}\mathbf{X}\rangle)^2.
\end{aligned}$$

This means we minimize variance of returns while maximizing expected return. This is in close resemblance with Markowitz-type portfolio selection. For a discussion of the relationship between Markowitz-type portfolio selection and the semi-log-optimal strategy, see [Ottucsák and Vajda (2007)] and Chapter 2 of this volume. The problem can also be formulated as a quadratic optimization problem:

$$\overline{\overline{f}}_{\mathbf{X}}(\mathbf{b}) = \langle\mathbf{b},\mathbf{R}\mathbf{b}\rangle + 4 - 4\langle\mathbf{b},\mathbf{m}\rangle$$

where

$$\mathbf{R} = \mathbb{E}(\mathbf{X}\mathbf{X}^T),$$

and

$$\mathbf{m} = \mathbb{E}(\mathbf{X}).$$

Note that $\mathbf{R}$ is symmetric and positive semidefinite, since, for any $\mathbf{z} \in \mathbb{R}^d$,

$$\mathbf{z}^T \mathbf{R} \mathbf{z} = \mathbf{z}^T \mathbb{E}(\mathbf{X} \mathbf{X}^T) \mathbf{z} = \mathbb{E}(\mathbf{z}^T \mathbf{X})^2 \geq 0.$$

This means we again face a convex programming problem.

4.3. Short Selling

4.3.1. *No-Ruin Constraints*

Short selling an asset is usually done by borrowing the asset under consideration and selling it. As collateral the investor has to provide securities of the same value to the lender of the shorted asset. This ensures that if anything goes wrong, the lender still has a high recovery rate.

While the investor has to provide collateral, after selling the borrowed assets, he again obtains the price of the shorted asset. This means that short selling is virtually for free:

$$S' = S - C + P,$$

where S' is the wealth after opening the short position, S is the wealth before, C is the collateral for borrowing and P is the price income of selling the asset being shorted. For simplicity, we assume

$$C = P,$$

hence,

$$S' = S,$$

and short selling is free. In practice, the act of short selling is more complicated. For institutional investors, the size of collateral depends on supply and demand on the short market, and the receiver of the more liquid asset usually pays interest. For simplicity, we ignore these problems.

Let us elaborate this process using a real-life example. Assume the investor wants to short sell 10 shares of IBM at $100, and he has $1000 in cash. First he has to find a lender – the short provider – who is willing to lend the shares. After exchanging the shares and the $1000 collateral, the investor sells the borrowed shares. After selling, the investor again has $1000 in cash, and the obligation to cover the shorted assets later.

In contrast with our modelling approach where short selling is free, it is also modeled in the literature such that selling an asset short yields immediate cash – this is called a naked short transaction. This is the case in the chapter on Mean-Variance Portfolio Theory of [Luenberger (1998)] and in [Cover and Ordentlich (1998)].

Assume our only investment is in asset j and our initial wealth is S_0. We invest a proportion of $b \in (-1,1)$ of our wealth. If the position is long ($b > 0$), it results in wealth

$$S_0(1-b) + S_0 b x_1^{(j)} = S_0 + S_0 b \left(x_1^{(j)} - 1 \right),$$

while, if the position is short ($b < 0$), we win as much money as the price drop of the asset:

$$S_0 + S_0|b|\left(1 - x_1^{(j)}\right) = S_0 + S_0 b \left(x_1^{(j)} - 1 \right).$$

In line with the previous example, assume that our investor has shorted 10 shares of IBM, at \$100. If the price drops \$10, he has to cover the short position at \$90, thus he gains $10 \cdot \$10$. If the price rises \$10, he has to cover at \$110, loosing $10 \cdot \$10$.

Let $\mathbf{b} = (b^{(0)}, b^{(1)}, \dots, b^{(d)})$ be the portfolio vector such that the 0-th component corresponds to cash. At the end of the first trading period, the investor's wealth becomes

$$S_1 = S_0 \left(b^{(0)} + \sum_{j=1}^{d} \left[b^{(j)^+} x_1^{(j)} + b^{(j)^-} (x_1^{(j)} - 1) \right] \right)^+, \qquad (4.9)$$

where $(\cdot)^+$ and $(\cdot)^-$ denote the positive and negative part operation. In the case of the investor's net wealth falling to zero or below he defaults. Note that negative wealth is not allowed in our framework by definition. Since only long positions cost money in this case, we will confine ourselves to portfolios such that $\sum_{j=0}^{d} b^{(j)^+} = 1$. Considering this, it is also true that

$$S_1 = S_0 \left(\sum_{j=0}^{d} b^{(j)^+} + \sum_{j=1}^{d} \left[b^{(j)^+}(x_1^{(j)} - 1) + b^{(j)^-}(x_1^{(j)} - 1) \right] \right)^+$$

$$= S_0 \left(1 + \sum_{j=1}^{d} \left[b^{(j)}(x_1^{(j)} - 1) \right] \right)^+. \qquad (4.10)$$

This shows that we gain as much as long positions raise and short positions fall.

We can see that short selling is a risky investment, because it is possible to default on total initial wealth without the default of any of the assets in the portfolio. The possibility of this would lead to a growth rate of minus infinity, thus we restrict our market according to

$$1 - B + \delta < x_n^{(j)} < 1 + B - \delta \qquad j = 1, \ldots, d. \tag{4.11}$$

Besides aiming at no-ruin, the role of $\delta > 0$ is ensuring that rate of growth is finite for any portfolio vector (i.e. $> -\infty$).

For the usual stock market daily data, there exist $0 < a_1 < 1 < a_2 < \infty$ such that

$$a_1 \leq x_n^{(j)} \leq a_2$$

for all $j = 1, \ldots, d$, for example, $a_1 = 0.7$ with $a_2 = 1.2$ (cf. [Fernholz (2000)]). Thus, we can choose $B = 0.3$.

Given (4.10) and (4.11) it is easy to see that maximum loss that we could suffer is $B \sum_{j=1}^{d} |b^{(j)}|$. This value has to be constrained to ensure no-ruin.

We denote the set of possible portfolio vectors by

$$
\Delta_d^{(-B)}
$$
$$
= \left\{ \mathbf{b} = (b^{(0)}, b^{(1)}, \ldots, b^{(d)}); \ b^{(0)} \geq 0, \sum_{j=0}^{d} b^{(j)^+} = 1, \ B \sum_{j=1}^{d} |b^{(j)}| \leq 1 \right\}.
$$
$$\tag{4.12}$$

$\sum_{j=0}^{d} b^{(j)^+} = 1$ means that we invest all of our initial wealth into some assets – buying long – or cash. By $B \sum_{j=1}^{d} |b^{(j)}| \leq 1$, maximum exposure is limited such that ruin is not possible, and the rate of growth is finite. $b^{(0)}$ is not included in the latter inequality, since possessing cash does not pose a risk. Notice that, if $B \leq 1$, then $\Delta_{d+1} \subset \Delta_d^{(-B)}$, and, so, the achievable growth rate with short selling can not be smaller than in the long-only case.

According to (4.10) and (4.11) with $B \le 1$, we show that ruin is impossible:

$$1 + \sum_{j=1}^{d} \left[b^{(j)}(x_1^{(j)} - 1) \right]$$

$$> 1 + \sum_{j=1}^{d} \left[b^{(j)+}(1 - B + \delta - 1) + b^{(j)-}(1 + B - \delta - 1) \right]$$

$$= 1 - (B - \delta) \sum_{j=1}^{d} |b^{(j)}|$$

$$\ge \delta \sum_{j=1}^{d} |b^{(j)}|.$$

If $\sum_{j=1}^{d} |b^{(j)}| = 0$ then $b^{(0)} = 1$, hence no-ruin. In any other case, $\delta \sum_{j=1}^{d} |b^{(j)}| > 0$, hence we have not only ensured no-ruin, but also

$$\mathbb{E} \ln \left(1 + \sum_{j=1}^{d} \left[b^{(j)}(X_1^{(j)} - 1) \right] \right)^+ > -\infty.$$

4.3.2. *Optimality Condition for Short Selling with Cash Account*

A problem with $\Delta_d^{(-B)}$ is its non-convexity. To see this, consider

$$\mathbf{b}_1 = (0, 1) \in \Delta_1^{(-1)},$$

$$\mathbf{b}_2 = (1, -1/2) \in \Delta_1^{(-1)},$$

with

$$\frac{\mathbf{b}_1 + \mathbf{b}_2}{2} = (1/2, 1/4) \notin \Delta_1^{(-1)}.$$

This means we can not simply apply the Kuhn–Tucker theorem on $\Delta_d^{(-B)}$.

Given a cash balance, we can transform our non-convex $\Delta_d^{(-B)}$ to a convex region $\widetilde{\Delta}_d^{(-B)}$, where application of our tools established in long-only investment becomes feasible. The new set of possible portfolio vectors is a convex region:

$$\widetilde{\Delta}_d^{(-B)} = \left\{ \widetilde{\mathbf{b}} = (\widetilde{b}^{(0+)}, \widetilde{b}^{(1+)}, \widetilde{b}^{(1-)}, \dots, \widetilde{b}^{(d+)}, \widetilde{b}^{(d-)}) \in \mathbb{R}_0^{+2d+1}; \right.$$

$$\left. \sum_{j=0}^{d} \widetilde{b}^{(j+)} = 1, B \sum_{j=1}^{d} (\widetilde{b}^{(j+)} + \widetilde{b}^{(j-)}) \le 1 \right\}.$$

Mapping from $\Delta_d^{(-B)}$ to $\widetilde{\Delta}_d^{(-B)}$ happens by

$$\widetilde{\mathbf{b}} = (b^{(0)}, (b^{(1)})^+, |(b^{(1)})^-|, \ldots, (b^{(d)})^+, |(b^{(d)})^-|). \qquad (4.13)$$

(4.9) implies that

$$S_1 = S_0 \left(\widetilde{b}^{(0+)} + \sum_{j=1}^{d} \left[\widetilde{b}^{(j+)} x_1^{(j)} + \widetilde{b}^{(j-)} (1 - x_1^{(j)}) \right] \right)^+,$$

thus, in line with the portfolio vector being transformed, we transform the market vector to

$$\widetilde{\mathbf{X}} = (1, X^{(1)}, 1 - X^{(1)}, \ldots, X^{(d)}, 1 - X^{(d)}),$$

so that

$$S_1 = S_0 \left\langle \widetilde{\mathbf{b}}, \widetilde{\mathbf{X}} \right\rangle.$$

To use the Kuhn–Tucker theorem, we enumerate linear, inequality-type constraints over the search space

$$B \sum_{j=1}^{d} (\widetilde{b}^{(j+)} + \widetilde{b}^{(j-)}) \leq 1,$$

and

$$\widetilde{b}^{(0+)} \geq 0, \widetilde{b}^{(i+)} \geq 0, \widetilde{b}^{(i-)} \geq 0,$$

for $i = 1, \ldots, d$. Our only equality-type constraint is

$$\sum_{j=0}^{d} \widetilde{b}^{(j+)} = 1.$$

The partial derivatives of the convex objective function $f_{\mathbf{X}}(\widetilde{\mathbf{b}}) = -\mathbb{E}\ln\left\langle \widetilde{\mathbf{b}}, \widetilde{\mathbf{X}} \right\rangle$ are

$$\frac{\partial f_{\mathbf{X}}(\widetilde{\mathbf{b}})}{\partial \widetilde{b}^{(0+)}} = -\mathbb{E}\frac{1}{\left\langle \widetilde{\mathbf{b}}, \widetilde{\mathbf{X}} \right\rangle},$$

$$\frac{\partial f_{\mathbf{X}}(\widetilde{\mathbf{b}})}{\partial \widetilde{b}^{(i+)}} = -\mathbb{E}\frac{X^{(i)}}{\left\langle \widetilde{\mathbf{b}}, \widetilde{\mathbf{X}} \right\rangle},$$

$$\frac{\partial f_{\mathbf{X}}(\widetilde{\mathbf{b}})}{\partial \widetilde{b}^{(i-)}} = -\mathbb{E}\frac{1 - X^{(i)}}{\left\langle \widetilde{\mathbf{b}}, \widetilde{\mathbf{X}} \right\rangle},$$

for $i = 1, \ldots, d$.

According to the Kuhn–Tucker theorem (KT), the portfolio vector $\widetilde{\mathbf{b}}^*$ is optimal if and only if there are KT multipliers assigned to each of the former $2d + 3$ constraints $\mu_{0_+} \geq 0, \mu_{i_+} \geq 0, \mu_{i_-} \geq 0, \nu_B \geq 0$ $(i = 1, \ldots, d)$ and $\vartheta \in \mathbb{R}$, such that

$$-\mathbb{E}\frac{1}{\left\langle \widetilde{\mathbf{b}}^*, \mathbf{X} \right\rangle} + \vartheta - \mu_{0_+} = 0, \tag{4.14}$$

$$-\mathbb{E}\frac{X^{(i)}}{\left\langle \widetilde{\mathbf{b}}^*, \mathbf{X} \right\rangle} + \vartheta - \mu_{i_+} + \nu_B B = 0,$$

$$-\mathbb{E}\frac{1 - X^{(i)}}{\left\langle \widetilde{\mathbf{b}}^*, \mathbf{X} \right\rangle} - \mu_{i_-} + \nu_B B = 0,$$

and

$$\mu_{0_+}\widetilde{b}^{*(0_+)} = 0,$$
$$\mu_{i_+}\widetilde{b}^{*(i_+)} = 0,$$
$$\mu_{i_-}\widetilde{b}^{*(i_-)} = 0,$$

for $i = 1, \ldots, d$, while

$$\nu_B \left[B \sum_{j=1}^{d} (\widetilde{b}^{(j_+)} + \widetilde{b}^{(j_-)}) - 1 \right] = 0, \tag{4.15}$$

$$\nu_B B \sum_{j=1}^{d} (\widetilde{b}^{(j_+)} + \widetilde{b}^{(j_-)}) = \nu_B.$$

Summing up equations in (4.14) weighted by $\widetilde{b}^{*(0)}, \widetilde{b}^{*(i_+)}, \widetilde{b}^{*(i_-)}$, we obtain

$$-\mathbb{E}\frac{\left\langle \widetilde{\mathbf{b}}^*, \widetilde{\mathbf{X}} \right\rangle}{\left\langle \widetilde{\mathbf{b}}^*, \widetilde{\mathbf{X}} \right\rangle} + \vartheta \sum_{j=0}^{d} \widetilde{b}^{*(j_+)} + \nu_B B \sum_{j=1}^{d} (\widetilde{b}^{(j_+)} + \widetilde{b}^{(j_-)}) = 0,$$

$$-1 + \vartheta + \nu_B = 0,$$
$$\vartheta = 1 - \nu_B, \tag{4.16}$$
$$\vartheta \leq 1.$$

In the case of $B \sum_{j=1}^{d} (\widetilde{b}^{(j_+)} + \widetilde{b}^{(j_-)}) < 1$, because of (4.15) and (4.16) we have that

$$\nu_B = 0, \text{ hence } \vartheta = 1.$$

This implies

$$-\mathbb{E}\frac{1}{\left\langle \widetilde{\mathbf{b}}^*, \widetilde{\mathbf{X}} \right\rangle} + 1 - \mu_{0_+} = 0,$$

$$-\mathbb{E}\frac{X^{(i)}}{\left\langle \widetilde{\mathbf{b}}^*, \widetilde{\mathbf{X}} \right\rangle} + 1 - \mu_{i_+} = 0,$$

$$-\mathbb{E}\frac{1 - X^{(i)}}{\left\langle \widetilde{\mathbf{b}}^*, \widetilde{\mathbf{X}} \right\rangle} - \mu_{i_-} = 0.$$

These equations result in the following additional properties

$$\widetilde{b}^{*(0_+)} > 0 \implies \mu_{0_+} = 0 \implies \mathbb{E}\frac{1}{\left\langle \widetilde{\mathbf{b}}^*, \widetilde{\mathbf{X}} \right\rangle} = 1,$$

$$\widetilde{b}^{*(0_+)} = 0 \implies \mu_{0_+} \geq 0 \implies \mathbb{E}\frac{1}{\left\langle \widetilde{\mathbf{b}}^*, \widetilde{\mathbf{X}} \right\rangle} \leq 1,$$

and

$$\widetilde{b}^{*(i_+)} > 0 \implies \mu_{i_+} = 0 \implies \mathbb{E}\frac{X^{(i)}}{\left\langle \widetilde{\mathbf{b}}^*, \widetilde{\mathbf{X}} \right\rangle} = 1,$$

$$\widetilde{b}^{*(i_+)} = 0 \implies \mu_{i_+} \geq 0 \implies \mathbb{E}\frac{X^{(i)}}{\left\langle \widetilde{\mathbf{b}}^*, \widetilde{\mathbf{X}} \right\rangle} \leq 1,$$

and

$$\widetilde{b}^{*(i_-)} > 0 \implies \mu_{i_-} = 0 \implies \mathbb{E}\frac{1 - X^{(i)}}{\left\langle \widetilde{\mathbf{b}}^*, \widetilde{\mathbf{X}} \right\rangle} = 0,$$

$$\widetilde{b}^{*(i_-)} = 0 \implies \mu_{i_-} \geq 0 \implies \mathbb{E}\frac{1 - X^{(i)}}{\left\langle \widetilde{\mathbf{b}}^*, \widetilde{\mathbf{X}} \right\rangle} \leq 0,$$

for $i = 1, \ldots, d$.

We transform the vector $\widetilde{\mathbf{b}}^*$ to the vector $\mathbf{b}^*$ such that

$$b^{*(i)} = \widetilde{b}^{*(i_+)} - \widetilde{b}^{*(i_-)}$$

$(i = 1, \ldots, d)$, while

$$b^{*(0)} = \widetilde{b}^{*(0)} + \sum_{i=1}^{d} \min\{\widetilde{b}^{*(i_+)}, \widetilde{b}^{*(i_-)}\}.$$

This way

$$\sum_{j=1}^{d} |b^{*(j)}| = 1,$$

and we have the same market exposure with $\mathbf{b}^*$ as with $\widetilde{\mathbf{b}}$.

Due to the simple mapping (4.13), with regard to the original portfolio vector $\mathbf{b}^*$, this means

$$b^{*(0)} > 0 \Longrightarrow \mathbb{E}\frac{1}{\left\langle \widetilde{\mathbf{b}}^*, \widetilde{\mathbf{X}} \right\rangle} = 1, \tag{4.17}$$

$$b^{*(0)} = 0 \Longrightarrow \mathbb{E}\frac{1}{\left\langle \widetilde{\mathbf{b}}^*, \widetilde{\mathbf{X}} \right\rangle} \leq 1.$$

Also

$$b^{*(i)} > 0 \Longrightarrow \mathbb{E}\frac{X^{(i)}}{\left\langle \widetilde{\mathbf{b}}^*, \widetilde{\mathbf{X}} \right\rangle} = 1 \text{ and } \mathbb{E}\frac{1 - X^{(i)}}{\left\langle \widetilde{\mathbf{b}}^*, \widetilde{\mathbf{X}} \right\rangle} \leq 0,$$

which is equivalent to

$$\mathbb{E}\frac{1}{\left\langle \widetilde{\mathbf{b}}^*, \widetilde{\mathbf{X}} \right\rangle} \leq \mathbb{E}\frac{X^{(i)}}{\left\langle \widetilde{\mathbf{b}}^*, \widetilde{\mathbf{X}} \right\rangle} = 1,$$

and

$$b^{*(i)} = 0 \Longrightarrow \mathbb{E}\frac{X^{(i)}}{\left\langle \widetilde{\mathbf{b}}^*, \widetilde{\mathbf{X}} \right\rangle} \leq 1 \text{ and } \mathbb{E}\frac{1 - X^{(i)}}{\left\langle \widetilde{\mathbf{b}}^*, \widetilde{\mathbf{X}} \right\rangle} \leq 0,$$

which is equivalent to

$$\mathbb{E}\frac{1}{\left\langle \widetilde{\mathbf{b}}^*, \widetilde{\mathbf{X}} \right\rangle} \leq \mathbb{E}\frac{X^{(i)}}{\left\langle \widetilde{\mathbf{b}}^*, \widetilde{\mathbf{X}} \right\rangle} \leq 1,$$

and

$$b^{*(i)} < 0 \Longrightarrow \mathbb{E}\frac{X^{(i)}}{\left\langle \widetilde{\mathbf{b}}^*, \widetilde{\mathbf{X}} \right\rangle} \leq 1 \text{ and } \mathbb{E}\frac{1 - X^{(i)}}{\left\langle \widetilde{\mathbf{b}}^*, \widetilde{\mathbf{X}} \right\rangle} = 0,$$

which is equivalent to

$$\mathbb{E}\frac{1}{\left\langle \widetilde{\mathbf{b}}^*, \widetilde{\mathbf{X}} \right\rangle} \leq \mathbb{E}\frac{X^{(i)}}{\left\langle \widetilde{\mathbf{b}}^*, \widetilde{\mathbf{X}} \right\rangle} \leq 1,$$

for $i = 1, \dots, d$.

4.4. Long-Only Leveraged Investment

4.4.1. *No-Ruin Condition*

In the leveraged frameworks we assume (4.11), thus market exposure can be increased over one without the possibility of ruin. Again, we denote the portfolio vector by

$$\mathbf{b} = (b^{(0)}, b^{(1)}, \dots, b^{(d)}),$$

where $b^{(0)} \geq 0$ stands for the cash balance, and, since there is no short-selling, $b^{(i)} \geq 0, i = 1, \dots, d$.

Assume the investor can borrow money and invest it on the same rate r. Assume also that the maximum investable amount of cash $L_{B,r}$ (relative to the initial wealth S_0) is always available the investor. In what follows, we refer to $L_{B,r}$ as buying power. $L_{B,r}$ is the maximum investable amount such that ruin does not occur, and is given by (4.11). Because our investor decides over the distribution of his buying power,

$$\sum_{j=0}^{d} b^{(j)} = L_{B,r}.$$

Unspent cash earns the same interest, r, as the rate of lending. The market vector is defined as

$$\mathbf{X}_r = (X^{(0)}, X^{(1)}, \dots, X^{(d)}) = (1 + r, X^{(1)}, \dots, X^{(d)}),$$

so $X^{(0)} = 1 + r$. The feasible set of portfolio vectors is

$$^r\Delta_d^{+B} = \left\{ \mathbf{b} = (b^{(0)}, b^{(1)}, \dots, b^{(d)}) \in \mathbb{R}_0^{+d+1}, \sum_{j=0}^{d} b^{(j)} = L_{B,r} \right\},$$

where $b^{(0)}$ denotes unspent buying power. The market evolves according to

$$S_1 = S_0(\langle \mathbf{b}, \mathbf{X}_r \rangle - (L_{B,r} - 1)(1 + r))^+,$$

where $S_0 r(L_{B,r} - 1)$ is the interest on borrowing $L_{B,r} - 1$ times the initial wealth S_0.

To ensure no-ruin conditions and finiteness of growth rate, choose

$$L_{B,r} = \frac{1 + r}{B + r}. \tag{4.18}$$

This ensures that ruin is not possible:

$$\langle \mathbf{b}, \mathbf{X}_r \rangle - (L_{B,r} - 1)(1 + r)$$

$$= \sum_{j=0}^{d} b^{(j)} X^{(j)} - (L_{B,r} - 1)(1 + r)$$

$$= b^{(0)}(1 + r) + \sum_{j=1}^{d} b^{(j)} X^{(j)} - (L_{B,r} - 1)(1 + r)$$

$$> b^{(0)}(1 + r) + \sum_{j=1}^{d} b^{(j)}(1 - B + \delta) - (L_{B,r} - 1)(1 + r)$$

$$= b^{(0)}(1 + r) + (L_{B,r} - b^{(0)})(1 - B + \delta) - (L_{B,r} - 1)(1 + r)$$

$$= b^{(0)}(r + B - \delta) - L_{B,r}(B - \delta + r) + 1 + r$$

$$\geq -\frac{1 + r}{B + r}(B - \delta + r) + 1 + r$$

$$= \delta \frac{1 + r}{B + r}.$$

4.4.2. *Kuhn–Tucker Characterization*

Our convex objective function, equivalent to the negative asymptotic growth rate, is

$$f_{\mathbf{X}_r}^{+B}(\mathbf{b}) = -\mathbb{E}\ln(\langle \mathbf{b}, \mathbf{X}_r \rangle - (L_{B,r} - 1)(1 + r)).$$

The linear inequality-type constraints are

$$-b^{(i)} \leq 0,$$

for $i = 0, \ldots, d$, while our only equality-type constraint is

$$\sum_{j=0}^{d} b^{(j)} - L_{B,r} = 0.$$

The partial derivatives of the optimized function are

$$\frac{\partial f_{\mathbf{X}_r}^{+B}(\mathbf{b})}{\partial b^{(i)}} = -\mathbb{E}\frac{X^{(i)}}{\langle \mathbf{b}, \mathbf{X}_r \rangle - (L_{B,r} - 1)(1 + r)}.$$

According to the Kuhn–Tucker necessary and sufficient theorem, a portfolio vector $\mathbf{b}^*$ is optimal if and only if there are KT multipliers $\mu_j \geq 0$ $(j = 0, \ldots, d)$ and $\vartheta \in \mathbb{R}$, such that

$$-\mathbb{E}\frac{X^{(j)}}{\langle \mathbf{b}^*, \mathbf{X}_r \rangle - (L_{B,r} - 1)(1 + r)} - \mu_j + \vartheta = 0 \qquad (4.19)$$

and

$$\mu_j b^{*(j)} = 0,$$

for $j = 0, \ldots, d$. Summing up (4.19) weighted by $b^{*(j)}$ we obtain

$$-\mathbb{E}\frac{\langle \mathbf{b}^*, \mathbf{X}_r \rangle}{\langle \mathbf{b}^*, \mathbf{X}_r \rangle - (L_{B,r} - 1)(1 + r)} - \sum_{j=0}^{d} \mu_j b^{*(j)} + \sum_{j=0}^{d} b^{*(j)} \vartheta = 0,$$

$$1 + \mathbb{E}\frac{(L_{B,r} - 1)(1 + r)}{\langle \mathbf{b}^*, \mathbf{X}_r \rangle - (L_{B,r} - 1)(1 + r)} = L_{B,r}\vartheta,$$

$$\frac{1}{L_{B,r}} + \frac{(L_{B,r} - 1)(1 + r)}{L_{B,r}}\mathbb{E}\frac{1}{\langle \mathbf{b}^*, \mathbf{X}_r \rangle - (L_{B,r} - 1)(1 + r)} = \vartheta. \quad (4.20)$$

This means that

$$b^{*(j)} > 0 \implies \mu_j = 0 \implies \mathbb{E}\frac{X^{(j)}}{\langle \mathbf{b}^*, \mathbf{X}_r \rangle - (L_{B,r} - 1)(1 + r)} = \vartheta, \quad (4.21)$$

and

$$b^{*(j)} = 0 \implies \mathbb{E}\frac{X^{(j)}}{\langle \mathbf{b}^*, \mathbf{X}_r \rangle - (L_{B,r} - 1)(1 + r)} \le \vartheta.$$

For the cash account, this means

$$b^{*(0)} > 0 \implies \mu_j = 0 \implies \mathbb{E}\frac{1 + r}{\langle \mathbf{b}^*, \mathbf{X}_r \rangle - (L_{B,r} - 1)(1 + r)} = \vartheta, \quad (4.22)$$

and

$$b^{*(0)} = 0 \implies \mathbb{E}\frac{1 + r}{\langle \mathbf{b}^*, \mathbf{X}_r \rangle - (L_{B,r} - 1)(1 + r)} \le \vartheta.$$

4.5. Short Selling and Leverage

For this case, we need to use both techniques of the previous sections. The market evolves according to

$$S_1 = S_0 \Big(b^{(0)}(1 + r)$$

$$+ \sum_{j=1}^{d} \Big[b^{(j)^+} x_1^{(j)} + b^{(j)^-}(x_1^{(j)} - 1 - r) \Big] - (L_{B,r} - 1)(1 + r) \Big)^+,$$

over the non-convex region

$$^r\Delta_d^{\pm B} = \Big\{ \mathbf{b} = (b^{(0)}, b^{(1)}, b^{(2)}, \ldots, b^{(d)}); \sum_{j=0}^{d} |b^{(j)}| = L_{B,r} \Big\},$$

where $L_{B,r}$ is the buying power defined in (4.18) and $b^{(0)}$ denotes unspent buying power. Again, one can check that the choice of $L_{B,r}$ ensures no-ruin conditions and finiteness of growth rate.

With the help of our technique developed in the short selling framework, we convert to the convex region

$$^r\widetilde{\Delta}_d^{\pm B} = \left\{ \widetilde{\mathbf{b}} = (\widetilde{b}^{(0+)}, \widetilde{b}^{(1+)}, \widetilde{b}^{(1-)}, \dots, \widetilde{b}^{(d+)}, \widetilde{b}^{(d-)}) \in \mathbb{R}_0^{+\,2d+1}; \right.$$

$$\left. \widetilde{b}^{(0+)} + \sum_{j=1}^d \left(\widetilde{b}^{(j+)} + \widetilde{b}^{(j-)} \right) = L_{B,r} \right\},$$

such that

$$\widetilde{\mathbf{b}} = (\widetilde{b}^{(0)}, \widetilde{b}^{(1_+)}, \widetilde{b}^{(1_-)} \dots, \widetilde{b}^{(d_+)}, \widetilde{b}^{(d_-)})$$
$$= (b^{(0)}, b^{(1)^+}, |b^{(1)^-}|, \dots, b^{(d)^+}, |b^{(d)^-}|).$$

Similarly to the short-selling case, we introduce the transformed return vector. Given

$$\mathbf{X} = (X^{(1)}, \dots, X^{(d)}),$$

we introduce

$$\mathbf{X}_{\pm r} = (1+r, X^{(1)}, 2 - X^{(1)} + r, \dots, X^{(d)}, 2 - X^{(d)} + r).$$

We introduce r in $2 - X^{(i)} + r$ terms, since short selling is free, hence, buying power spent on short positions still earns interest. We use $2 - X^{(i)} + r$ instead of $1 - X^{(i)} + r$, since, while short selling is actually free, it still limits our buying power, which is the basis of the convex formulation.

Because of $^r\widetilde{\Delta}_d^{\pm B} = {}^r\Delta_{2d}^{+B}$, we can easily apply (4.21) and (4.22), hence

$$b^{*(0)} > 0 \Longrightarrow \mathbb{E} \frac{1+r}{\left\langle \widetilde{\mathbf{b}}^*, \mathbf{X}_{\pm r} \right\rangle - (L_{B,r} - 1)(1+r)} = \vartheta,$$

$$b^{*(0)} = 0 \Longrightarrow \mathbb{E} \frac{1+r}{\left\langle \widetilde{\mathbf{b}}^*, \mathbf{X}_{\pm r} \right\rangle - (L_{B,r} - 1)(1+r)} \leq \vartheta,$$

where ϑ is defined by (4.20) with $\mathbf{X}_r = \mathbf{X}_{\pm r}$ in place, and

$$b^{*(i)} > 0 \Longrightarrow$$

$$\mathbb{E} \frac{X^{(i)}}{\left\langle \widetilde{\mathbf{b}}^*, \mathbf{X}_{\pm r} \right\rangle - (L_{B,r} - 1)(1+r)} = \vartheta,$$

$$\mathbb{E} \frac{2 - X^{(i)} + r}{\left\langle \widetilde{\mathbf{b}}^*, \mathbf{X}_{\pm r} \right\rangle - (L_{B,r} - 1)(1+r)} \leq \vartheta$$

and

$$b^{*(i)} = 0 \implies$$

$$\mathbb{E} \frac{X^{(i)}}{\left\langle \widetilde{\mathbf{b}}^*, \mathbf{X}_{\pm r} \right\rangle - (L_{B,r} - 1)(1 + r)} \leq \vartheta,$$

$$\mathbb{E} \frac{2 - X^{(i)} + r}{\left\langle \widetilde{\mathbf{b}}^*, \mathbf{X}_{\pm r} \right\rangle - (L_{B,r} - 1)(1 + r)} \leq \vartheta$$

and

$$b^{*(i)} < 0 \implies$$

$$\mathbb{E} \frac{X^{(i)}}{\left\langle \widetilde{\mathbf{b}}^*, \mathbf{X}_{\pm r} \right\rangle - (L_{B,r} - 1)(1 + r)} \leq \vartheta,$$

$$\mathbb{E} \frac{2 - X^{(i)} + r}{\left\langle \widetilde{\mathbf{b}}^*, \mathbf{X}_{\pm r} \right\rangle - (L_{B,r} - 1)(1 + r)} = \vartheta.$$

Note that, in the special case of $L_{B,r} = 1$, we have $\vartheta = 1$ because of (4.20).

4.6. Experiments

Our empirical investigation considers three situations, each of which is considered in long-only, short, leveraged and leveraged-short cases. We examine the BCRP strategy, which chooses the best constant-portfolio vector with hindsight, and its empirical causal counterpart, the causal i.i.d. strategy. The latter strategy uses the best portfolio based on past data, and it is asymptotically optimal for i.i.d. returns. The third algorithm is asymptotically optimal for the case of Markovian time series. Using the nearest-neighbor-based portfolio selection (cf. Chapter 2 of this volume) with 100 neighbors, we investigate whether shorting yields extra growth in the case of dependent market returns.

The New York Stock Exchange (NYSE) data set [Gelencsér and Ottucsák (2006)] includes daily closing prices of 19 assets along a 44-year period ending in 2006. The same data is used in Chapter 2 of this volume, which facilitates comparison of algorithms.

Interest rate is constant in our experiments. We calculated effective daily yield over the 44 years based on the Federal Reserve Fund Rate from the FRED database. The annual rate in this period is 6.3%, which is equivalent to $r = 0.000245$ daily interest.

Regarding (4.11), we chose the conservative bound $B = 0.4$, as the largest one-day change in asset value over the 44 years has been 0.3029. This bound implies that, in the case of $r = 0$, the maximum leverage is $L_{B,r} = 2.5$-fold, while, in the case of $r = 0.000245$, $L_{B,r} = 2.4991$. Performance of BCRP algorithms improves further by decreasing B until $B = 0.2$, but this limit would not guarantee a no-ruin situation. This property also implies that optimal leverage factor on our dataset is less than 5.

Given our convex formalism for the space of the portfolio vector and the convexity of log utility, we use Lagrange multipliers and active-set algorithms for the optimization.

Table 4.1 shows the results of the BCRP experiments. Shorting does not have any effect in this case, while leverage results in significant gain. The behavior of shorting strategies is in line with intuition, since taking a permanently short position of an asset is not beneficial. The leveraged strategies use maximum leverage, and they do not only increase market exposure, but invest into more assets in order to reduce variation of the portfolio. This is in contrast to the behavior of leveraged mean-variance optimal portfolios.

Table 4.1. Average Annual Yields and optimal portfolios on NYSE data.

Asset	AAY	b^*	b^*_{-B}	b^*_{+B}	$b^*_{\pm B}$
Cash/Debt	–	0	0	-1.4991	-1.4991
AHP	13%	0	0	0	0
ALCOA	9%	0	0	0	0
AMERB	14%	0	0	0.01	0.01
COKE	14%	0	0	0	0
DOW	12%	0	0	0	0
DUPONT	9%	0	0	0	0
FORD	9%	0	0	0	0
GE	13%	0	0	0	0
GM	7%	0	0	0	0
HP	15%	0.17	0.17	0.32	0.32
IBM	10%	0	0	0	0
INGER	11%	0	0	0	0
JNJ	16%	0	0	0.48	0.48
KIMBC	13%	0	0	0.03	0.03
MERCK	15%	0	0	0.14	0.14
MMM	11%	0	0	0	0
MORRIS	20%	0.75	0.75	1.15	1.15
PANDG	13%	0	0	0	0
SCHLUM	15%	0.08	0.08	0.36	0.36
AAY		20%	20%	34%	34%

Table 4.2.   Average Annual Yields.

Strategy		Annual Average Yield
BCRP	Long only	20%
	Short	20%
	Leverage	34%
	Short & Leverage	34%
IID	Long only	13%
	Short	11%
	Leverage	16%
	Short & Leverage	14%
Nearest Neighbor	Long only	32%
	Short	30%
	Leverage	66%
	Short & Leverage	83%

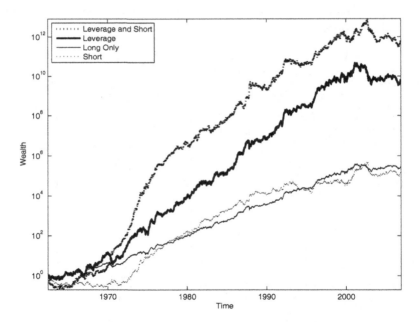

Fig. 4.1.   Cumulative wealth of the nearest-neighbor strategy starting from 1962.

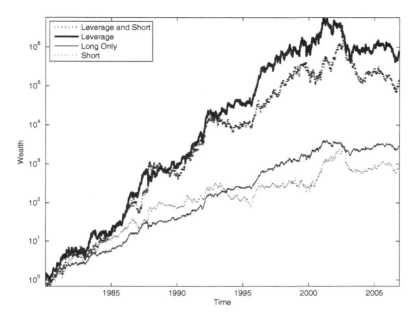

Fig. 4.2. Cumulative wealth of the nearest-neighbor strategy starting from 1980.

Table 4.2 presents growth rates of the three cases we consider. BCRP being an optimistically-anticipating estimate of possible growth, our i.i.d. strategies do significantly underperform, while the average annual yields (AAYs) of the nearest-neighbor strategies including leverage are spectacular. Figure 4.1 presents the evolution of wealth in the latter case. While allowing short positions results in large drawdowns in the beginning, these algorithms catch up later. Figure 4.2 shows the result of the algorithms starting from 1980; it reveals that short selling does not offer any significant gain in this period compared with the long-only approaches.

References

Bell, R. M. and Cover, T. M. (1980). Competitive optimality of logarithmic investment, *Mathematics of Operations Research* **5**, pp. 161–166.

Breiman, L. (1961). Optimal gambling systems for favorable games, in *Proceedings of the Fourth Berkeley Symposium on Mathematical Statistics and Probability* (University of California Press, Berkeley), pp. 65–78.

Cover, T. M. (1984). An algorithm for maximizing expected log investment return, *IEEE Transactions on Information Theory* **30**, pp. 369–373.

Cover, T. M. and Ordentlich, E. (1998). Universal portfolios with short sales and margin, in *Proceedings of IEEE International Symposium on Information Theory* (Cambridge, MA), pp. 174–174.

Cover, T. M. and Thomas, J. A. (1991). *Elements of Information Theory* (Wiley, New York).

Fernholz, E. R. (2000). *Stochastic Portfolio Theory* (Springer, New York).

Finkelstein, M. and Whitley, R. (1981). Optimal strategies for repeated games, *Advances in Applied Probability* **13**, pp. 415–428.

Gelencsér, G. and Ottucsák, G. (2006). NYSE data sets at the log-optimal portfolio homepage, URL www.cs.bme.hu/~oti/portfolio.

Györfi, L., Lugosi, G. and Udina, F. (2006). Nonparametric kernel based sequential investment strategies, *Mathematical Finance* **16**, pp. 337–357.

Kelly, J. L. (1956). A new interpretation of information rate, *Bell System Technical Journal* **35**, pp. 917–926.

Kuhn, H. W. and Tucker, A. W. (1951). Nonlinear programming, in *Proceedings of 2nd Berkeley Symposium* (University of California Press, Berkeley), pp. 481–492.

Latané, H. A. (1959). Criteria for choice among risky ventures, *Journal of Political Economy* **38**, pp. 145–155.

Luenberger, D. G. (1998). *Investment Science* (Oxford University Press, Oxford).

Markowitz, H. (1952). Portfolio selection, *Journal of Finance* **7**, 1, pp. 77–91.

Markowitz, H. (1976). Investment for the long run: New evidence for an old rule, *Journal of Finance* **31**, 5, pp. 1273–1286.

Móri, T. F. (1982). On favourable stochastic games, *Annales Univ. Sci. Budapest. R. Eötvös Nom., Sect. Comput* **3**, pp. 99–103.

Móri, T. F. and Székely, G. J. (1982). How to win if you can? in *Proceedings of the Colloquia Mathematica Societatis János Bolyai 36. Limit Theorems in Probability and Statistics*, pp. 791–806.

Ottucsák, G. and Vajda, I. (2007). An asymptotic analysis of the mean-variance portfolio selection, *Statistics & Decisions* **25**, pp. 63–88.

Pulley, L. B. (1983). Mean-variance approximation to expected logarithmic utility, *Operations Research* **IT-40**, pp. 685–696.

Roll, R. (1973). Evidence on the "growth-optimum" model, *The Journal of Finance* **28**, pp. 551–566.

Vajda, I. (2006). Analysis of semi-log-optimal investment strategies, in M. Huskova and M. Janzura (eds.), *Prague Stochastics 2006* (MATFYZ-PRESS, Prague).

Chapter 5

Nonparametric Sequential Prediction of Stationary Time Series

László Györfi[*] and György Ottucsák[†]

*Department of Computer Science and Information Theory,
Budapest University of Technology and Economics,
H-1117, Magyar tudósok körútja 2., Budapest, Hungary.*
[*] gyorfi@cs.bme.hu, [†] oti@cs.bme.hu

We present simple procedures for the prediction of a real-valued time series with side information. For squared loss (regression problem), we survey the basic principles of universally-consistent estimates. The prediction algorithms are based on a combination of several simple predictors. We show that, if the sequence is a realization of a stationary and ergodic random process, then the average of squared errors converges, almost surely, to that of the optimum, given by the Bayes predictor. We offer an analog result for the prediction of stationary Gaussian processes.

5.1. Introduction

We study the problem of sequential prediction of a real-valued sequence. At each time instant $t = 1, 2, \ldots$, the predictor is asked to guess the value of the next outcome y_t of a sequence of real numbers $y_1, y_2, \ldots$ with knowledge of the past $y_1^{t-1} = (y_1, \ldots, y_{t-1})$ (where y_1^0 denotes the empty string) and the side information vectors $x_1^t = (x_1, \ldots, x_t)$, where $x_t \in \mathbb{R}^d$. Thus, the predictor's estimate, at time t, is based on the value of x_1^t and y_1^{t-1}. A prediction strategy is a sequence $g = \{g_t\}_{t=1}^\infty$ of functions

$$g_t : \left(\mathbb{R}^d\right)^t \times \mathbb{R}^{t-1} \to \mathbb{R}$$

so that the prediction formed at time t is $g_t(x_1^t, y_1^{t-1})$.

In this study, we assume that $(x_1, y_1), (x_2, y_2), \ldots$ are realizations of the random variables $(X_1, Y_1), (X_2, Y_2), \ldots$, such that $\{(X_n, Y_n)\}_{-\infty}^\infty$ is a jointly stationary and ergodic process.

After n time instants, the *normalized cumulative prediction error* is

$$L_n(g) = \frac{1}{n} \sum_{t=1}^{n} (g_t(X_1^t, Y_1^{t-1}) - Y_t)^2.$$

Our aim is to achieve small $L_n(g)$ when n is large.

For this prediction problem, an example can be the forecast of daily relative prices y_t of an asset, while the side information vector x_t may contain some information on other assets in the past days or the trading volume in the previous day or some news related to the actual assets, etc. This is a widely-investigated research problem. However, in the vast majority of the corresponding literature, the side information is not included in the model. Moreover, a parametric model (AR, MA, ARMA, ARIMA, ARCH, GARCH, etc.) is fitted to the stochastic process $\{Y_t\}$, its parameters are estimated, and a prediction is derived from the parameter estimates (cf. [Tsay (2002)]). Formally, this approach means that there is a parameter, θ, such that the best predictor has the form

$$\mathbb{E}\{Y_t \mid Y_1^{t-1}\} = g_t(\theta, Y_1^{t-1}),$$

for a function g_t. The parameter θ is estimated from the past data Y_1^{t-1}, and the estimate is denoted by $\hat{\theta}$. Then the data-driven predictor is

$$g_t(\hat{\theta}, Y_1^{t-1}).$$

Here we do not assume any parametric model, so our results are fully nonparametric. This modelling is important for financial data when the process is only approximately governed by stochastic differential equations, so the parametric modelling can be weak. Moreover the error criterion of the parameter estimate (usually the maximum likelihood estimate) has no relation to the mean square error of the prediction derived. The main aim of this research is to construct predictors, called *universally-consistent predictors*, which are consistent for all stationary time series. Such universal feature can be proven using the recent principles of nonparametric statistics and machine learning algorithms.

The results below are given in an autoregressive framework, that is, the value Y_t is predicted based on X_1^t and Y_1^{t-1}. The fundamental limit for the predictability of the sequence can be determined based on a result of [Algoet (1994)], who showed that for any prediction strategy g and stationary ergodic process $\{(X_n, Y_n)\}_{-\infty}^{\infty}$,

$$\liminf_{n \to \infty} L_n(g) \geq L^* \quad \text{almost surely}, \tag{5.1}$$

where

$$L^* = \mathbb{E}\left\{ \left(Y_0 - \mathbb{E}\{Y_0 | X_{-\infty}^0, Y_{-\infty}^{-1}\} \right)^2 \right\}$$

is the minimal mean-squared error of any prediction for the value of Y_0 based on the infinite past $X_{-\infty}^0, Y_{-\infty}^{-1}$. Note that it follows by stationarity and the martingale convergence theorem (see, e.g., [Stout (1974)]) that

$$L^* = \lim_{n \to \infty} \mathbb{E}\left\{ \left(Y_n - \mathbb{E}\{Y_n | X_1^n, Y_1^{n-1}\} \right)^2 \right\}.$$

This lower bound gives sense to the following definition:

Definition 5.1. A prediction strategy g is called *universally consistent with respect to a class $\mathcal{C}$ of stationary and ergodic processes* $\{(X_n, Y_n)\}_{-\infty}^{\infty}$, if, for each process in the class,

$$\lim_{n \to \infty} L_n(g) = L^* \quad \text{almost surely.}$$

Universally-consistent strategies asymptotically achieve the best possible squared loss for all ergodic processes in the class. [Algoet (1992)] and [Morvai *et al.* (1996)] proved that there exists a prediction strategy universal with respect to the class of all bounded ergodic processes. However, the prediction strategies exhibited in these papers are either very complex or have an unreasonably slow rate of convergence even for well-behaved processes.

Next, we introduce several simple prediction strategies which, apart from having the above-mentioned universal property of [Algoet (1992)] and [Morvai *et al.* (1996)], promise much improved performance for "nice" processes. The algorithms build on a methodology worked out in recent years for prediction of individual sequences, see [Vovk (1990)], [Feder *et al.* (1992)], [Littlestone and Warmuth (1994)], [Cesa-Bianchi *et al.* (1997)], [Merhav and Feder (1998)], [Kivinen and Warmuth (1999)], [Singer and Feder (1999)], [Cesa-Bianchi and Lugosi (2006)] for a survey.

An approach similar to the one of this paper was adopted by [Györfi *et al.* (1999)], where prediction of stationary binary sequences was addressed. There they introduced a simple randomized predictor which predicts asymptotically as well as the optimal predictor for all binary ergodic processes. The present setup and results differ in several important points from those of [Györfi *et al.* (1999)]. On the one hand, special properties of the squared loss function considered here allow us to avoid randomization of the predictor, and to define a significantly simpler prediction scheme. On the other hand, possible unboundedness of a real-valued process requires special care, which we demonstrate on the example of Gaussian processes. We refer to [Singer and Feder (1999, 2000)], [Yang (2000)], [Nobel (2003)], for recent closely related work.

In Section 5.2 we survey the basic principles of nonparametric regression estimates. In Section 5.3 we introduce universally-consistent strategies for bounded ergodic processes which are based on a combination of partitioning or kernel or nearest-neighbor or generalized linear estimates. In Section 5.4 we consider the prediction of unbounded sequences including the ergodic Gaussian process.

5.2. Nonparametric Regression Estimation

5.2.1. *The Regression Problem*

For the prediction of time series, an important source of the basic principles is the nonparametric regression. In regression analysis, one considers a random vector (X, Y), where X is $\mathbb{R}^d$-valued and Y is $\mathbb{R}$-valued, and one is interested how the value of the so-called response variable Y depends on the value of the observation vector X. This means that one wants to find a function $f : \mathbb{R}^d \to \mathbb{R}$, such that $f(X)$ is a "good approximation of Y", that is, $f(X)$ should be close to Y in some sense, which is equivalent to making $|f(X) - Y|$ "small". Since X and Y are random vectors, $|f(X) - Y|$ is random as well, therefore it is not clear what "small $|f(X) - Y|$" means. We can resolve this problem by introducing the so-called L_2 *risk* or *mean-squared error* of f,

$$\mathbb{E}|f(X) - Y|^2,$$

and requiring it to be as small as possible.

Therefore, we are interested in a function $m^* : \mathbb{R}^d \to \mathbb{R}$ such that

$$\mathbb{E}|m^*(X) - Y|^2 = \min_{f:\mathbb{R}^d \to \mathbb{R}} \mathbb{E}|f(X) - Y|^2.$$

Such a function can be obtained explicitly as follows. Let

$$m(x) = \mathbb{E}\{Y|X = x\}$$

be the *regression function*. We will show that the regression function minimizes the L_2 risk. Indeed, for an arbitrary $f : \mathbb{R}^d \to \mathbb{R}$, a version of the Steiner theorem implies that

$$\mathbb{E}|f(X) - Y|^2 = \mathbb{E}|f(X) - m(X) + m(X) - Y|^2$$
$$= \mathbb{E}|f(X) - m(X)|^2 + \mathbb{E}|m(X) - Y|^2,$$

where we have used

$$
\begin{aligned}
&\mathbb{E}\left\{(f(X) - m(X))(m(X) - Y)\right\} \\
&= \mathbb{E}\left\{\mathbb{E}\left\{(f(X) - m(X))(m(X) - Y)|X\right\}\right\} \\
&= \mathbb{E}\left\{(f(X) - m(X))\mathbb{E}\{m(X) - Y|X\}\right\} \\
&= \mathbb{E}\left\{(f(X) - m(X))(m(X) - m(X))\right\} \\
&= 0.
\end{aligned}
$$

Hence,

$$
\mathbb{E}|f(X) - Y|^2 = \int_{\mathbb{R}^d} |f(x) - m(x)|^2 \mu(dx) + \mathbb{E}|m(X) - Y|^2, \qquad (5.2)
$$

where μ denotes the distribution of X. The first term is called the L_2 error of f. It is always nonnegative and is zero if $f(x) = m(x)$. Therefore, $m^*(x) = m(x)$, i.e., the optimal approximation (with respect to the L_2 risk) of Y by a function of X is given by $m(X)$.

5.2.2. *Regression Function Estimation and L_2 Error*

In applications the distribution of (X, Y) (and hence also the regression function) is usually unknown. Therefore it is impossible to predict Y using $m(X)$. But it is often possible to observe data according to the distribution of (X, Y) and to estimate the regression function from these data.

To be more precise, denote by (X, Y), (X_1, Y_1), $(X_2, Y_2), \ldots$ independent and identically distributed (i.i.d.) random variables with $\mathbb{E}Y^2 < \infty$. Let $\mathcal{D}_n$ be the set of *data* defined by

$$
\mathcal{D}_n = \{(X_1, Y_1), \ldots, (X_n, Y_n)\}.
$$

In the regression function estimation problem one wants to use the data $\mathcal{D}_n$ in order to construct an estimate $m_n : \mathbb{R}^d \to \mathbb{R}$ of the regression function m. Here $m_n(x) = m_n(x, \mathcal{D}_n)$ is a measurable function of x and the data. For simplicity, we will suppress $\mathcal{D}_n$ in the notation and write $m_n(x)$ instead of $m_n(x, \mathcal{D}_n)$.

In general, estimates will not be equal to the regression function. To compare different estimates, we need an error criterion which measures the difference between the regression function and an arbitrary estimate m_n. One of the key points we would like to make is that the motivation for introducing the regression function leads naturally to an L_2 error criterion for measuring the performance of the regression function estimate. Recall that the main goal was to find a function f such that the L_2 risk

$\mathbb{E}|f(X) - Y|^2$ is small. The minimal value of this L_2 risk is $\mathbb{E}|m(X) - Y|^2$, and it is achieved by the regression function m. Similarly to (5.2), one can show that the L_2 risk $\mathbb{E}\{|m_n(X) - Y|^2|\mathcal{D}_n\}$ of an estimate m_n satisfies

$$\mathbb{E}\left\{|m_n(X) - Y|^2|\mathcal{D}_n\right\} = \int_{\mathbb{R}^d} |m_n(x) - m(x)|^2 \mu(dx) + \mathbb{E}|m(X) - Y|^2. \quad (5.3)$$

Thus, the L_2 risk of an estimate m_n is close to the optimal value if and only if the L_2 error

$$\int_{\mathbb{R}^d} |m_n(x) - m(x)|^2 \mu(dx) \quad (5.4)$$

is close to zero. Therefore we will use the L_2 error (5.4) in order to measure the quality of an estimate and we will study estimates for which this L_2 error is small.

In this section we describe the basic principles of nonparametric regression estimation: *local averaging*, *local modelling*, *global modelling* (or *least-squares estimation*), and *penalized modelling*. (Concerning the details see [Győrfi *et al.* (2002)].)

Recall that the data can be written as

$$Y_i = m(X_i) + \epsilon_i,$$

where $\epsilon_i = Y_i - m(X_i)$ satisfies $\mathbb{E}(\epsilon_i|X_i) = 0$. Thus, Y_i can be considered as the sum of the value of the regression function at X_i and some error ϵ_i, where the expected value of the error is zero. This motivates the construction of the estimates by *local averaging*, i.e., estimation of $m(x)$ by the average of those Y_i where X_i is "close" to x. Such an estimate can be written as

$$m_n(x) = \sum_{i=1}^{n} W_{n,i}(x) \cdot Y_i,$$

where the weights $W_{n,i}(x) = W_{n,i}(x, X_1, \ldots, X_n) \in \mathbb{R}$ depend on $X_1, \ldots, X_n$. Usually the weights are nonnegative and $W_{n,i}(x)$ is "small" if X_i is "far" from x.

5.2.3. *Partitioning Estimate*

An example of such an estimate is the *partitioning estimate*. Here one chooses a finite or countably infinite partition $\mathcal{P}_n = \{A_{n,1}, A_{n,2}, \ldots\}$ of $\mathbb{R}^d$ consisting of cells $A_{n,j} \subseteq \mathbb{R}^d$ and defines, for $x \in A_{n,j}$, the estimate by averaging Y_i's with the corresponding X_i's in $A_{n,j}$, i.e.,

$$m_n(x) = \frac{\sum_{i=1}^n I_{\{X_i \in A_{n,j}\}} Y_i}{\sum_{i=1}^n I_{\{X_i \in A_{n,j}\}}} \quad \text{for } x \in A_{n,j}, \tag{5.5}$$

where I_A denotes the indicator function of set A, so

$$W_{n,i}(x) = \frac{I_{\{X_i \in A_{n,j}\}}}{\sum_{l=1}^n I_{\{X_l \in A_{n,j}\}}} \quad \text{for } x \in A_{n,j}.$$

Here and in the following we use the convention $\frac{0}{0} = 0$. In order to have consistency, on the one hand we need that the cells $A_{n,j}$ should be "small", and on the other hand the number of non-zero terms in the denominator of (5.5) should be "large". These requirements can be satisfied if the sequences of partition $\mathcal{P}_n$ are asymptotically fine, i.e., if

$$\text{diam}(A) = \sup_{x,y \in A} \|x - y\|$$

denotes the diameter of a set, then, for each sphere S centered at the origin,

$$\lim_{n \to \infty} \max_{j: A_{n,j} \cap S \neq \emptyset} \text{diam}(A_{n,j}) = 0$$

and

$$\lim_{n \to \infty} \frac{|\{j : A_{n,j} \cap S \neq \emptyset\}|}{n} = 0.$$

For the partition $\mathcal{P}_n$, the most important example is when the cells $A_{n,j}$ are cubes of volume h_n^d. For cubic partition, the consistency conditions above mean that

$$\lim_{n \to \infty} h_n = 0 \quad \text{and} \quad \lim_{n \to \infty} n h_n^d = \infty. \tag{5.6}$$

Next, we bound the rate of convergence of $\mathbb{E}\|m_n - m\|^2$ for cubic partitions and regression functions which are Lipschitz continuous.

Proposition 5.1. *For a cubic partition with side length h_n, assume that*

$$\mathbf{Var}(Y|X = x) \leq \sigma^2, \ x \in \mathbb{R}^d,$$

$$|m(x) - m(z)| \leq C\|x - z\|, \ x, z \in \mathbb{R}^d, \tag{5.7}$$

and that X has a compact support S. Then

$$\mathbb{E}\|m_n - m\|^2 \leq \frac{c_1}{n \cdot h_n^d} + d \cdot C^2 \cdot h_n^2,$$

thus, for

$$h_n = c_2 n^{-\frac{1}{d+2}}$$

we get

$$\mathbb{E}\|m_n - m\|^2 \le c_3 n^{-2/(d+2)}.$$

In order to prove Proposition 5.1 we need the following technical lemma. An integer-valued random variable $B(n,p)$ is said to be binomially distributed with parameters n and $0 \le p \le 1$ if

$$\mathbb{P}\{B(n,p) = k\} = \binom{n}{k} p^k (1-p)^{n-k}, \quad k = 0, 1, \dots, n.$$

Lemma 5.1. *Let the random variable $B(n,p)$ be binomially distributed with parameters n and p. Then:*

(i)

$$\mathbb{E}\left\{\frac{1}{1+B(n,p)}\right\} \le \frac{1}{(n+1)p},$$

(ii)

$$\mathbb{E}\left\{\frac{1}{B(n,p)} I_{\{B(n,p)>0\}}\right\} \le \frac{2}{(n+1)p}.$$

Proof. Part (i) follows from the following simple calculation:

$$\begin{aligned}
\mathbb{E}\left\{\frac{1}{1+B(n,p)}\right\} &= \sum_{k=0}^{n} \frac{1}{k+1} \binom{n}{k} p^k (1-p)^{n-k} \\
&= \frac{1}{(n+1)p} \sum_{k=0}^{n} \binom{n+1}{k+1} p^{k+1} (1-p)^{n-k} \\
&\le \frac{1}{(n+1)p} \sum_{k=0}^{n+1} \binom{n+1}{k} p^k (1-p)^{n-k+1} \\
&= \frac{1}{(n+1)p} (p + (1-p))^{n+1} \\
&= \frac{1}{(n+1)p}.
\end{aligned}$$

For (ii) we have

$$\mathbb{E}\left\{\frac{1}{B(n,p)} I_{\{B(n,p)>0\}}\right\} \le \mathbb{E}\left\{\frac{2}{1+B(n,p)}\right\} \le \frac{2}{(n+1)p}$$

by (i). $\qquad\square$

Proof of Proposition 5.1. If $A_n(x)$ denotes the cell of the partition into which x falls, then set the semi-estimate

$$\hat{m}_n(x) = \mathbb{E}\{m_n(x)|X_1,\ldots,X_n\} = \frac{\sum_{i=1}^{n} m(X_i)I_{\{X_i \in A_n(x)\}}}{n\mu_n(A_n(x))},$$

where μ_n denotes the empirical distribution for $X_1,\ldots,X_n$. Then

$$\mathbb{E}\{(m_n(x) - m(x))^2|X_1,\ldots,X_n\}$$
$$= \mathbb{E}\{(m_n(x) - \hat{m}_n(x))^2|X_1,\ldots,X_n\} + (\hat{m}_n(x) - m(x))^2. \quad (5.8)$$

We have

$$\mathbb{E}\{(m_n(x) - \hat{m}_n(x))^2|X_1,\ldots,X_n\}$$
$$= \mathbb{E}\left\{\left(\frac{\sum_{i=1}^{n}(Y_i - m(X_i))I_{\{X_i \in A_n(x)\}}}{n\mu_n(A_n(x))}\right)^2 \Big| X_1,\ldots,X_n\right\}$$
$$= \frac{\sum_{i=1}^{n} \mathbf{Var}(Y_i|X_i)I_{\{X_i \in A_n(x)\}}}{(n\mu_n(A_n(x)))^2}$$
$$\leq \frac{\sigma^2}{n\mu_n(A_n(x))}I_{\{n\mu_n(A_n(x))>0\}}.$$

By Jensen's inequality

$$(\hat{m}_n(x) - m(x))^2 = \left(\frac{\sum_{i=1}^{n}(m(X_i) - m(x))I_{\{X_i \in A_n(x)\}}}{n\mu_n(A_n(x))}\right)^2 I_{\{n\mu_n(A_n(x))>0\}}$$
$$+ m(x)^2 I_{\{n\mu_n(A_n(x))=0\}}$$
$$\leq \frac{\sum_{i=1}^{n}(m(X_i) - m(x))^2 I_{\{X_i \in A_n(x)\}}}{n\mu_n(A_n(x))}I_{\{n\mu_n(A_n(x))>0\}}$$
$$+ m(x)^2 I_{\{n\mu_n(A_n(x))=0\}}$$
$$\leq d \cdot C^2 h_n^2 I_{\{n\mu_n(A_n(x))>0\}} + m(x)^2 I_{\{n\mu_n(A_n(x))=0\}}$$
$$\text{(by (5.7) and } \max_{z \in A_n(x)} \|x - z\| \leq d \cdot h_n^2)$$
$$\leq d \cdot C^2 h_n^2 + m(x)^2 I_{\{n\mu_n(A_n(x))=0\}}.$$

Without loss of generality, assume that S is a cube and the union of $A_{n,1},\ldots,A_{n,l_n}$ is S. Then

$$l_n \leq \frac{\tilde{c}}{h_n^d}$$

for some constant $\tilde{c}$ proportional to the volume of S and, by Lemma 5.1 and (5.8),

$$
\mathbb{E}\left\{\int (m_n(x) - m(x))^2 \mu(dx)\right\}
$$
$$
= \mathbb{E}\left\{\int (m_n(x) - \hat{m}_n(x))^2 \mu(dx)\right\} + \mathbb{E}\left\{\int (\hat{m}_n(x) - m(x))^2 \mu(dx)\right\}
$$
$$
= \sum_{j=1}^{l_n} \mathbb{E}\left\{\int_{A_{n,j}} (m_n(x) - \hat{m}_n(x))^2 \mu(dx)\right\}
$$
$$
+ \sum_{j=1}^{l_n} \mathbb{E}\left\{\int_{A_{n,j}} (\hat{m}_n(x) - m(x))^2 \mu(dx)\right\}
$$
$$
\leq \sum_{j=1}^{l_n} \mathbb{E}\left\{\frac{\sigma^2 \mu(A_{n,j})}{n\mu_n(A_{n,j})} I_{\{n\mu_n(A_{n,j})>0\}}\right\} + dC^2 h_n^2
$$
$$
+ \sum_{j=1}^{l_n} \mathbb{E}\left\{\int_{A_{n,j}} m(x)^2 \mu(dx) I_{\{\mu_n(A_{n,j})=0\}}\right\}
$$
$$
\leq \sum_{j=1}^{l_n} \frac{2\sigma^2 \mu(A_{n,j})}{n\mu(A_{n,j})} + dC^2 h_n^2 + \sum_{j=1}^{l_n} \int_{A_{n,j}} m(x)^2 \mu(dx) \mathbb{P}\{\mu_n(A_{n,j}) = 0\}
$$
$$
\leq l_n \frac{2\sigma^2}{n} + dC^2 h_n^2 + \sup_{z \in S}\{m(z)^2\} \sum_{j=1}^{l_n} \mu(A_{n,j})(1 - \mu(A_{n,j}))^n
$$
$$
\leq l_n \frac{2\sigma^2}{n} + dC^2 h_n^2 + l_n \frac{\sup_{z \in S} m(z)^2}{n} \sup_j n\mu(A_{n,j}) e^{-n\mu(A_{n,j})}
$$
$$
\leq l_n \frac{2\sigma^2}{n} + dC^2 h_n^2 + l_n \frac{\sup_{z \in S} m(z)^2 e^{-1}}{n}
$$
$$
\text{(since } \sup_u u e^{-u} = e^{-1})
$$
$$
\leq \frac{(2\sigma^2 + \sup_{z \in S} m(z)^2 e^{-1})\tilde{c}}{n h_n^d} + dC^2 h_n^2.
$$

$\square$

5.2.4. Kernel Estimate

The second example of a local averaging estimate is the *Nadaraya–Watson kernel estimate*. Let $K : \mathbb{R}^d \to \mathbb{R}_+$ be a function called the kernel function, and let $h > 0$ be a bandwidth. The kernel estimate is defined by

$$m_n(x) = \frac{\sum_{i=1}^n K\left(\frac{x-X_i}{h}\right) Y_i}{\sum_{i=1}^n K\left(\frac{x-X_i}{h}\right)}, \tag{5.9}$$

so

$$W_{n,i}(x) = \frac{K\left(\frac{x-X_i}{h}\right)}{\sum_{j=1}^n K\left(\frac{x-X_j}{h}\right)}.$$

Here, the estimate is a weighted average of the Y_i, where the weight of Y_i (i.e., the influence of Y_i on the value of the estimate at x) depends on the distance between X_i and x. For the bandwidth $h = h_n$, the consistency conditions are (5.6). If one uses the so-called naïve kernel (or window kernel) $K(x) = I_{\{\|x\| \leq 1\}}$, then

$$m_n(x) = \frac{\sum_{i=1}^n I_{\{\|x-X_i\| \leq h\}} Y_i}{\sum_{i=1}^n I_{\{\|x-X_i\| \leq h\}}},$$

i.e., one estimates $m(x)$ by averaging Y_i's such that the distance between X_i and x is not greater than h.

In the sequel we bound the rate of convergence of $\mathbb{E}\|m_n - m\|^2$ for a naïve kernel and a Lipschitz continuous regression function.

Proposition 5.2. *For a kernel estimate with a naïve kernel assume that*

$$\mathbf{Var}(Y|X = x) \leq \sigma^2, \, x \in \mathbb{R}^d,$$

and

$$|m(x) - m(z)| \leq C\|x - z\|, \, x, z \in \mathbb{R}^d,$$

and X has a compact support S^*. Then

$$\mathbb{E}\|m_n - m\|^2 \leq \frac{c_1}{n \cdot h_n^d} + C^2 h_n^2,$$

thus, for

$$h_n = c_2 n^{-\frac{1}{d+2}}$$

we have

$$\mathbb{E}\|m_n - m\|^2 \leq c_3 n^{-2/(d+2)}.$$

Proof. We proceed similarly to Proposition 5.1. Put

$$\hat{m}_n(x) = \frac{\sum_{i=1}^n m(X_i) I_{\{X_i \in S_{x,h_n}\}}}{n \mu_n(S_{x,h_n})},$$

where $S_{x,h}$ denotes the sphere centered at x with radius h, then we have the decomposition (5.8). If $B_n(x) = \{n\mu_n(S_{x,h_n}) > 0\}$, then

$$\mathbb{E}\{(m_n(x) - \hat{m}_n(x))^2 | X_1, \ldots, X_n\}$$

$$= \mathbb{E}\left\{ \left(\frac{\sum_{i=1}^n (Y_i - m(X_i))I_{\{X_i \in S_{x,h_n}\}}}{n\mu_n(S_{x,h_n})} \right)^2 | X_1, \ldots, X_n \right\}$$

$$= \frac{\sum_{i=1}^n \mathbf{Var}(Y_i | X_i) I_{\{X_i \in S_{x,h_n}\}}}{(n\mu_n(S_{x,h_n}))^2}$$

$$\leq \frac{\sigma^2}{n\mu_n(S_{x,h_n})} I_{B_n(x)}.$$

By Jensen's inequality and the Lipschitz property of m,

$$(\hat{m}_n(x) - m(x))^2$$

$$= \left(\frac{\sum_{i=1}^n (m(X_i) - m(x))I_{\{X_i \in S_{x,h_n}\}}}{n\mu_n(S_{x,h_n})} \right)^2 I_{B_n(x)} + m(x)^2 I_{B_n(x)^c}$$

$$\leq \frac{\sum_{i=1}^n (m(X_i) - m(x))^2 I_{\{X_i \in S_{x,h_n}\}}}{n\mu_n(S_{x,h_n})} I_{B_n(x)} + m(x)^2 I_{B_n(x)^c}$$

$$\leq C^2 h_n^2 I_{B_n(x)} + m(x)^2 I_{B_n(x)^c}$$

$$\leq C^2 h_n^2 + m(x)^2 I_{B_n(x)^c}.$$

Using this, together with Lemma 5.1,

$$\mathbb{E}\left\{ \int (m_n(x) - m(x))^2 \mu(dx) \right\}$$

$$= \mathbb{E}\left\{ \int (m_n(x) - \hat{m}_n(x))^2 \mu(dx) \right\} + \mathbb{E}\left\{ \int (\hat{m}_n(x) - m(x))^2 \mu(dx) \right\}$$

$$\leq \int_{S^*} \mathbb{E}\left\{ \frac{\sigma^2}{n\mu_n(S_{x,h_n})} I_{\{\mu_n(S_{x,h_n}) > 0\}} \right\} \mu(dx) + C^2 h_n^2$$

$$+ \int_{S^*} \mathbb{E}\left\{ m(x)^2 I_{\{\mu_n(S_{x,h_n}) = 0\}} \right\} \mu(dx)$$

$$\leq \int_{S^*} \frac{2\sigma^2}{n\mu(S_{x,h_n})} \mu(dx) + C^2 h_n^2 + \int_{S^*} m(x)^2 (1 - \mu(S_{x,h_n}))^n \mu(dx)$$

$$\leq \int_{S^*} \frac{2\sigma^2}{n\mu(S_{x,h_n})} \mu(dx) + C^2 h_n^2 + \sup_{z \in S^*} m(z)^2 \int_{S^*} e^{-n\mu(S_{x,h_n})} \mu(dx)$$

$$\leq 2\sigma^2 \int_{S^*} \frac{1}{n\mu(S_{x,h_n})} \mu(dx) + C^2 h_n^2$$

$$+ \sup_{z \in S^*} m(z)^2 \max_u u e^{-u} \int_{S^*} \frac{1}{n\mu(S_{x,h_n})} \mu(dx).$$

We can find $z_1, \ldots, z_{M_n}$ such that the union of $S_{z_1, rh_n/2}, \ldots, S_{z_{M_n}, rh_n/2}$ covers S^*, and

$$M_n \leq \frac{\tilde{c}}{h_n^d}.$$

Then

$$\int_{S^*} \frac{1}{n\mu(S_{x, rh_n})} \mu(dx) \leq \sum_{j=1}^{M_n} \int \frac{I_{\{x \in S_{z_j, rh_n/2}\}}}{n\mu(S_{x, rh_n})} \mu(dx)$$

$$\leq \sum_{j=1}^{M_n} \int \frac{I_{\{x \in S_{z_j, rh_n/2}\}}}{n\mu(S_{z_j, rh_n/2})} \mu(dx)$$

$$\leq \frac{M_n}{n}$$

$$\leq \frac{\tilde{c}}{nh_n^d}.$$

Combining these inequalities completes the proof. □

5.2.5. Nearest-Neighbor Estimate

Our final example of local averaging estimates is the *k-nearest-neighbor* (*k*-NN) *estimate*. Here one determines the k nearest X_i's to x in terms of distance $\|x - X_i\|$ and estimates $m(x)$ by the average of the corresponding Y_i's. More precisely, for $x \in \mathbb{R}^d$, let

$$(X_{(1)}(x), Y_{(1)}(x)), \ldots, (X_{(n)}(x), Y_{(n)}(x))$$

be a permutation of

$$(X_1, Y_1), \ldots, (X_n, Y_n)$$

such that

$$\|x - X_{(1)}(x)\| \leq \cdots \leq \|x - X_{(n)}(x)\|.$$

The k-NN estimate is defined by

$$m_n(x) = \frac{1}{k} \sum_{i=1}^{k} Y_{(i)}(x). \tag{5.10}$$

Here, the weight $W_{ni}(x)$ equals $1/k$ if X_i is among the k nearest-neighbors of x, and equals 0 otherwise. If $k = k_n \to \infty$ such that $k_n/n \to 0$ then the k-nearest-neighbor regression estimate is consistent.

Next, we bound the rate of convergence of $\mathbb{E}\|m_n - m\|^2$ for a k_n-nearest-neighbor estimate.

Proposition 5.3. *Assume that X is bounded,*

$$\sigma^2(x) = \mathbf{Var}(Y|X = x) \le \sigma^2 \quad (x \in \mathbb{R}^d)$$

and

$$|m(x) - m(z)| \le C\|x - z\| \quad (x, z \in \mathbb{R}^d).$$

Assume that $d \ge 3$. Let m_n be the k_n-NN estimate. Then

$$\mathbb{E}\|m_n - m\|^2 \le \frac{\sigma^2}{k_n} + c_1 \left(\frac{k_n}{n}\right)^{2/d},$$

thus, for $k_n = c_2 n^{\frac{2}{d+2}}$,

$$\mathbb{E}\|m_n - m\|^2 \le c_3 n^{-\frac{2}{d+2}}.$$

For the proof of Proposition 5.3 we need the rate of convergence of nearest-neighbor distances.

Lemma 5.2. *Assume that X is bounded. If $d \ge 3$, then*

$$\mathbb{E}\{\|X_{(1,n)}(X) - X\|^2\} \le \frac{\tilde{c}}{n^{2/d}}.$$

Proof. For fixed $\epsilon > 0$,

$$\mathbb{P}\{\|X_{(1,n)}(X) - X\| > \epsilon\} = \mathbb{E}\{(1 - \mu(S_{X,\epsilon}))^n\}.$$

Let $A_1, \ldots, A_{N(\epsilon)}$ be a cubic partition of the bounded support of μ such that the A_j's have diameter ϵ and

$$N(\epsilon) \le \frac{c}{\epsilon^d}.$$

If $x \in A_j$, then $A_j \subset S_{x,\epsilon}$, therefore

$$\mathbb{E}\{(1 - \mu(S_{X,\epsilon}))^n\} = \sum_{j=1}^{N(\epsilon)} \int_{A_j} (1 - \mu(S_{x,\epsilon}))^n \mu(dx)$$

$$\le \sum_{j=1}^{N(\epsilon)} \int_{A_j} (1 - \mu(A_j))^n \mu(dx)$$

$$= \sum_{j=1}^{N(\epsilon)} \mu(A_j)(1 - \mu(A_j))^n.$$

Obviously,

$$\sum_{j=1}^{N(\epsilon)} \mu(A_j)(1 - \mu(A_j))^n \leq \sum_{j=1}^{N(\epsilon)} \max_z z(1 - z)^n$$

$$\leq \sum_{j=1}^{N(\epsilon)} \max_z z e^{-nz}$$

$$= \frac{e^{-1} N(\epsilon)}{n}.$$

If L stands for the diameter of the support of μ, then

$$\mathbb{E}\{\|X_{(1,n)}(X) - X\|^2\} = \int_0^\infty \mathbb{P}\{\|X_{(1,n)}(X) - X\|^2 > \epsilon\} d\epsilon$$

$$= \int_0^{L^2} \mathbb{P}\{\|X_{(1,n)}(X) - X\| > \sqrt{\epsilon}\} d\epsilon$$

$$\leq \int_0^{L^2} \min\left\{1, \frac{e^{-1} N(\sqrt{\epsilon})}{n}\right\} d\epsilon$$

$$\leq \int_0^{L^2} \min\left\{1, \frac{c}{en} \epsilon^{-d/2}\right\} d\epsilon$$

$$= \int_0^{(c/(en))^{2/d}} 1 \, d\epsilon + \frac{c}{en} \int_{(c/(en))^{2/d}}^{L^2} \epsilon^{-d/2} d\epsilon$$

$$\leq \frac{\tilde{c}}{n^{2/d}}$$

for $d \geq 3$.   □

Proof of Proposition 5.3. We have the decomposition

$$\mathbb{E}\{(m_n(x) - m(x))^2\} = \mathbb{E}\{(m_n(x) - \mathbb{E}\{m_n(x)|X_1, \ldots, X_n\})^2\}$$

$$+ \mathbb{E}\{(\mathbb{E}\{m_n(x)|X_1, \ldots, X_n\} - m(x))^2\}$$

$$= I_1(x) + I_2(x).$$

The first term is easier:

$$I_1(x) = \mathbb{E}\left\{\left(\frac{1}{k_n} \sum_{i=1}^{k_n} (Y_{(i,n)}(x) - m(X_{(i,n)}(x)))\right)^2\right\}$$

$$= \mathbb{E}\left\{\frac{1}{k_n^2} \sum_{i=1}^{k_n} \sigma^2(X_{(i,n)}(x))\right\}$$

$$\leq \frac{\sigma^2}{k_n}.$$

For the second term

$$I_2(x) = \mathbb{E}\left\{\left(\frac{1}{k_n}\sum_{i=1}^{k_n}(m(X_{(i,n)}(x)) - m(x))\right)^2\right\}$$

$$\leq \mathbb{E}\left\{\left(\frac{1}{k_n}\sum_{i=1}^{k_n}|m(X_{(i,n)}(x)) - m(x)|\right)^2\right\}$$

$$\leq \mathbb{E}\left\{\left(\frac{1}{k_n}\sum_{i=1}^{k_n}C\|X_{(i,n)}(x) - x\|\right)^2\right\}.$$

Put $N = k_n\lfloor\frac{n}{k_n}\rfloor$. Split the data $X_1,\ldots,X_n$ into $k_n + 1$ segments such that the first k_n segments have length $\lfloor\frac{n}{k_n}\rfloor$, and let $\tilde{X}_j^x$ be the first nearest neighbor of x from the jth segment. Then $\tilde{X}_1^x, \ldots, \tilde{X}_{k_n}^x$ are k_n different elements of $\{X_1,\ldots,X_n\}$, which implies

$$\sum_{i=1}^{k_n}\|X_{(i,n)}(x) - x\| \leq \sum_{j=1}^{k_n}\|\tilde{X}_j^x - x\|,$$

therefore, by Jensen's inequality,

$$I_2(x) \leq C^2\mathbb{E}\left\{\left(\frac{1}{k_n}\sum_{j=1}^{k_n}\|\tilde{X}_j^x - x\|\right)^2\right\}$$

$$\leq C^2\frac{1}{k_n}\sum_{j=1}^{k_n}\mathbb{E}\left\{\|\tilde{X}_j^x - x\|^2\right\}$$

$$= C^2\mathbb{E}\left\{\|\tilde{X}_1^x - x\|^2\right\}$$

$$= C^2\mathbb{E}\left\{\|X_{(1,\lfloor\frac{n}{k_n}\rfloor)}(x) - x\|^2\right\}.$$

Thus, by Lemma 5.2,

$$\frac{1}{C^2}\left\lfloor\frac{n}{k_n}\right\rfloor^{2/d}\int I_2(x)\mu(dx) \leq \left\lfloor\frac{n}{k_n}\right\rfloor^{2/d}\mathbb{E}\left\{\|X_{(1,\lfloor\frac{n}{k_n}\rfloor)}(X) - X\|^2\right\}$$

$$\leq const.$$

$\square$

5.2.6. Empirical Error Minimization

A generalization of the partitioning estimate leads to *global modelling* or *least-squares estimates*. Let $\mathcal{P}_n = \{A_{n,1}, A_{n,2},\ldots\}$ be a partition of $\mathbb{R}^d$ and let $\mathcal{F}_n$ be the set of all piecewise constant functions with respect to that partition, i.e.,

$$\mathcal{F}_n = \left\{ \sum_j a_j I_{A_{n,j}} : a_j \in \mathbb{R} \right\}. \tag{5.11}$$

Then it is easy to see that the partitioning estimate (5.5) satisfies

$$m_n(\cdot) = \arg\min_{f \in \mathcal{F}_n} \left\{ \frac{1}{n} \sum_{i=1}^{n} |f(X_i) - Y_i|^2 \right\}. \tag{5.12}$$

Hence it minimizes the empirical L_2 risk

$$\frac{1}{n} \sum_{i=1}^{n} |f(X_i) - Y_i|^2 \tag{5.13}$$

over $\mathcal{F}_n$. Least-squares estimates are defined by minimizing the empirical L_2 risk over a general set of functions $\mathcal{F}_n$ (instead of (5.11)). Observe that it does not make sense to minimize (5.13) over all functions f, because this may lead to a function which interpolates the data and hence is not a reasonable estimate. Thus, one has to restrict the set of functions over which one minimizes the empirical L_2 risk. Examples of possible choices of the set $\mathcal{F}_n$ are sets of piecewise polynomials with respect to a partition $\mathcal{P}_n$, or sets of smooth piecewise polynomials (splines). The use of spline spaces ensures that the estimate is a smooth function. Important members of least-squares estimates are the generalized linear estimates. Let $\{\phi_j\}_{j=1}^{\infty}$ be real-valued functions defined on $\mathbb{R}^d$ and let $\mathcal{F}_n$ be defined by

$$\mathcal{F}_n = \left\{ f; f = \sum_{j=1}^{\ell_n} c_j \phi_j \right\}.$$

Then the generalized linear estimate is defined by

$$m_n(\cdot) = \arg\min_{f \in \mathcal{F}_n} \left\{ \frac{1}{n} \sum_{i=1}^{n} (f(X_i) - Y_i)^2 \right\}$$

$$= \arg\min_{c_1,\dots,c_{\ell_n}} \left\{ \frac{1}{n} \sum_{i=1}^{n} \left(\sum_{j=1}^{\ell_n} c_j \phi_j(X_i) - Y_i \right)^2 \right\}.$$

If the set

$$\left\{ \sum_{j=1}^{\ell} c_j \phi_j; \ (c_1, \dots, c_\ell), \ \ell = 1, 2, \dots \right\}$$

is dense in the set of continuous functions of d variables, $\ell_n \to \infty$ and $\ell_n/n \to 0$ then the generalized linear regression estimate defined above is consistent. For least-squares estimates, other examples are neural networks, radial basis functions and orthogonal series estimates.

Next, we bound the rate of convergence of empirical error minimization estimates.

Condition (sG). The error $\varepsilon \triangleq Y - m(X)$ is a sub-Gaussian random variable, that is, there exist constants $\lambda > 0$ and $\Lambda < \infty$ with

$$\mathbb{E}\left\{ \exp(\lambda \varepsilon^2)\big|\, X \right\} < \Lambda \quad \text{almost surely.}$$

Furthermore, define $\sigma^2 \triangleq \mathbb{E}\{\varepsilon^2\}$ and set $\lambda_0 = 4\Lambda/\lambda$.

Condition (C). The class $\mathcal{F}_n$ is totally bounded with respect to the supremum norm. For each $\delta > 0$, let $M(\delta)$ denote the δ-covering number of $\mathcal{F}$. This means that for every $\delta > 0$, there is a δ-cover $f_1, \ldots, f_M$ with $M = M(\delta)$ such that

$$\min_{1 \leq i \leq M} \sup_x |f_i(x) - f(x)| \leq \delta$$

for all $f \in \mathcal{F}_n$. In addition, assume that $\mathcal{F}_n$ is uniformly bounded by L, that is,

$$|f(x)| \leq L < \infty$$

for all $x \in \mathbb{R}$ and $f \in \mathcal{F}_n$.

Proposition 5.4. *Assume that conditions (C) and (sG) hold and*

$$|m(x)| \leq L < \infty.$$

Then, for the estimate m_n defined by (5.5) and for all $\delta_n > 0$, $n \geq 1$,

$$\mathbb{E}\left\{ (m_n(X) - m(X))^2 \right\}$$
$$\leq 2 \inf_{f \in \mathcal{F}_n} \mathbb{E}\{(f(X) - m(X))^2\}$$
$$+ (16L + 4\sigma)\delta_n + \left(16L^2 + 4\max\left\{ L\sqrt{2\lambda_0}, 8\lambda_0 \right\}\right) \frac{\log M(\delta_n)}{n}.$$

In the proof of this proposition we use the following lemma:

Lemma 5.3. [Wegkamp (1999)] *Let Z be a random variable with*

$$\mathbb{E}\{Z\} = 0 \text{ and } \mathbb{E}\left\{ \exp(\lambda Z^2) \right\} \leq A$$

for some constants $\lambda > 0$ and $A \geq 1$. Then

$$\mathbb{E}\{\exp(\beta Z)\} \leq \exp\left(\frac{2A\beta^2}{\lambda}\right)$$

holds for every $\beta \in \mathbb{R}$.

Proof. Since for all $t > 0$, $\mathbb{P}\{|Z| > t\} \leq A\exp(-\lambda t^2)$ holds by Markov inequality, we have for all integers $m \geq 2$,

$$\mathbb{E}\{|Z|^m\} = \int_0^\infty \mathbb{P}\{|Z|^m > t\}\,dt$$

$$\leq A\int_0^\infty \exp\left(-\lambda t^{2/m}\right)dt = A\lambda^{-m/2}\Gamma\left(\frac{m}{2}+1\right).$$

Note that $\Gamma^2(\frac{m}{2}+1) \leq \Gamma(m+1)$ by Cauchy–Schwarz. The following inequalities are now self-evident:

$$\mathbb{E}\{\exp(\beta Z)\} = 1 + \sum_{m=2}^\infty \frac{1}{m!}\mathbb{E}(\beta Z)^m$$

$$\leq 1 + \sum_{m=2}^\infty \frac{1}{m!}|\beta|^m\mathbb{E}|Z|^m$$

$$\leq 1 + A\sum_{m=2}^\infty \lambda^{-m/2}|\beta|^m\frac{\Gamma\left(\frac{m}{2}+1\right)}{\Gamma(m+1)}$$

$$\leq 1 + A\sum_{m=2}^\infty \lambda^{-m/2}|\beta|^m\frac{1}{\Gamma\left(\frac{m}{2}+1\right)}$$

$$= 1 + A\sum_{m=1}^\infty \left(\frac{\beta^2}{\lambda}\right)^m\frac{1}{\Gamma(m+1)}$$

$$+ A\sum_{m=1}^\infty \left(\frac{\beta^2}{\lambda}\right)^{m+\frac{1}{2}}\frac{1}{\Gamma\left(m+\frac{3}{2}\right)}$$

$$\leq 1 + A\sum_{m=1}^\infty \left(\frac{\beta^2}{\lambda}\right)^m\left(1+\left(\frac{\beta^2}{\lambda}\right)^{\frac{1}{2}}\right)\frac{1}{\Gamma(m+1)}.$$

Finally, invoke the inequality $1 + (1+\sqrt{x})(\exp(x)-1) \leq \exp(2x)$ for $x > 0$, to obtain the result. $\square$

Lemma 5.4. [Antos *et al.* (2005)] *Let X_{ij}, $i = 1,\ldots,n$, $j = 1,\ldots M$ be random variables such that for each fixed j, $X_{1j},\ldots,X_{nj}$ are independent* and identically distributed such that for each $s_0 \geq s > 0$

$$\mathbb{E}\{e^{sX_{ij}}\} \leq e^{s^2 \sigma_j^2}.$$

For $\delta_j > 0$, put

$$\vartheta = \min_{j \leq M} \frac{\delta_j}{\sigma_j^2}.$$

Then

$$\mathbb{E}\left\{ \max_{j \leq M} \left(\frac{1}{n} \sum_{i=1}^{n} X_{ij} - \delta_j \right) \right\} \leq \frac{\log M}{\min\{\vartheta, s_0\} n}. \tag{5.14}$$

If

$$\mathbb{E}\{X_{ij}\} = 0$$

and

$$|X_{ij}| \leq K,$$

then

$$\mathbb{E}\left\{ \max_{j \leq M} \left(\frac{1}{n} \sum_{i=1}^{n} X_{ij} - \delta_j \right) \right\} \leq \max\{1/\vartheta^*, K\} \frac{\log M}{n}, \tag{5.15}$$

where

$$\vartheta^* = \min_{j \leq M} \frac{\delta_j}{\mathbf{Var}(X_{ij})}.$$

Proof. For the notation

$$Y_j = \frac{1}{n} \sum_{i=1}^{n} X_{ij} - \delta_j$$

we have that for any $s_0 \geq s > 0$

$$\begin{aligned}
\mathbb{E}\{e^{snY_j}\} &= \mathbb{E}\{e^{sn(\frac{1}{n}\sum_{i=1}^{n} X_{ij} - \delta_j)}\} \\
&= e^{-sn\delta_j} \left(\mathbb{E}\{e^{sX_{1j}}\} \right)^n \\
&\leq e^{-sn\delta_j} e^{ns^2\sigma_j^2} \\
&\leq e^{-sn\vartheta\sigma_j^2 + s^2 n\sigma_j^2}.
\end{aligned}$$

Thus

$$e^{sn\mathbb{E}\{\max_{j \le M} Y_j\}} \le \mathbb{E}\{e^{sn \max_{j \le M} Y_j}\}$$
$$= \mathbb{E}\{\max_{j \le M} e^{snY_j}\}$$
$$\le \sum_{j \le M} \mathbb{E}\{e^{snY_j}\}$$
$$\le \sum_{j \le M} e^{-sn\sigma_j^2(\vartheta - s)}.$$

For $s = \min\{\vartheta, s_0\}$ it implies that

$$\mathbb{E}\{\max_{j \le M} Y_j\} \le \frac{1}{sn} \log \left(\sum_{j \le M} e^{-sn\sigma_j^2(\vartheta - s)} \right) \le \frac{\log M}{\min\{\vartheta, s_0\}n}.$$

In order to prove the second half of the lemma, notice that, for any $L > 0$ and $|x| \le L$ we have the inequality

$$e^x = 1 + x + x^2 \sum_{i=2}^{\infty} \frac{x^{i-2}}{i!}$$
$$\le 1 + x + x^2 \sum_{i=2}^{\infty} \frac{L^{i-2}}{i!}$$
$$= 1 + x + x^2 \frac{e^L - 1 - L}{L^2},$$

therefore $0 < s \le s_0 = L/K$ implies that $s|X_{ij}| \le L$, so

$$e^{sX_{ij}} \le 1 + sX_{ij} + (sX_{ij})^2 \frac{e^L - 1 - L}{L^2}.$$

Thus,

$$\mathbb{E}\{e^{sX_{ij}}\} \le 1 + s^2 \mathbf{Var}(X_{ij}) \frac{e^L - 1 - L}{L^2} \le e^{s^2 \mathbf{Var}(X_{ij}) \frac{e^L - 1 - L}{L^2}},$$

so (5.15) follows from (5.14). □

Proof of Proposition 5.4. This proof is due to [Györfi and Wegkamp (2008)]. Set

$$D(f) = \mathbb{E}\{(f(X) - Y)^2\}$$

and

$$\hat{D}(f) = \frac{1}{n} \sum_{i=1}^{n} (f(X_i) - Y_i)^2$$

and

$$\Delta_f(x) = (m(x) - f(x))^2$$

and define

$$R(\mathcal{F}_n) \triangleq \sup_{f \in \mathcal{F}_n} \left[D(f) - 2\widehat{D}(f) \right] \le R_1(\mathcal{F}_n) + R_2(\mathcal{F}_n),$$

where

$$R_1(\mathcal{F}_n) \triangleq \sup_{f \in \mathcal{F}_n} \left[\frac{2}{n} \sum_{i=1}^{n} \{\mathbb{E}\Delta_f(X_i) - \Delta_f(X_i)\} - \frac{1}{2}\mathbb{E}\{\Delta_f(X)\} \right]$$

and

$$R_2(\mathcal{F}_n) \triangleq \sup_{f \in \mathcal{F}_n} \left[\frac{4}{n} \sum_{i=1}^{n} \varepsilon_i(f(X_i) - m(X_i)) - \frac{1}{2}\mathbb{E}\{\Delta_f(X)\} \right],$$

with $\varepsilon_i \triangleq Y_i - m(X_i)$. By the definition of $R(\mathcal{F}_n)$ and m_n, we have for all $f \in \mathcal{F}_n$

$$\begin{aligned}
\mathbb{E}\left\{ (m_n(X) - m(X))^2 \mid \mathcal{D}_n \right\} &= \mathbb{E}\left\{ D(m_n) \mid \mathcal{D}_n \right\} - D(m) \\
&\le 2\{\widehat{D}(m_n) - \widehat{D}(m)\} + R(\mathcal{F}_n) \\
&\le 2\{\widehat{D}(f) - \widehat{D}(m)\} + R(\mathcal{F}_n) .
\end{aligned}$$

After taking expectations on both sides, we obtain

$$\mathbb{E}\left\{ (m_n(X) - m(X))^2 \right\} \le 2\mathbb{E}\left\{ (f(X) - m(X))^2 \right\} + \mathbb{E}\{R(\mathcal{F}_n)\}.$$

Let $\mathcal{F}_n'$ be a finite δ_n-covering net (with respect to the sup-norm) of $\mathcal{F}_n$ with $M(\delta_n) = |\mathcal{F}_n'|$. It means that for any $f \in \mathcal{F}_n$ there is an $f' \in \mathcal{F}_n'$ such that

$$\sup_x |f(x) - f'(x)| \le \delta_n,$$

which implies that

$$\begin{aligned}
| (m(X_i) - f(X_i))^2 &- (m(X_i) - f'(X_i))^2 | \\
&\le |f(X_i) - f'(X_i)| \cdot \left(|m(X_i) - f(X_i)| + |m(X_i) - f'(X_i)| \right) \\
&\le 4L|f(X_i) - f'(X_i)| \\
&\le 4L\delta_n,
\end{aligned}$$

and, by Cauchy–Schwarz inequality,

$$\begin{aligned}
\mathbb{E}\{|\varepsilon_i(m(X_i) - f(X_i)) &- \varepsilon_i(m(X_i) - f'(X_i))|\} \\
&\le \sqrt{\mathbb{E}\{\varepsilon_i^2\}}\sqrt{\mathbb{E}\{(f(X_i) - f'(X_i))^2\}} \\
&\le \sigma\delta_n.
\end{aligned}$$

Thus,

$$\mathbb{E}\{R(\mathcal{F}_n)\} \leq 2\delta_n(4L + \sigma) + \mathbb{E}\{R(\mathcal{F}_n')\},$$

and, therefore,

$$\mathbb{E}\left\{(m_n(X) - m(X))^2\right\}$$
$$\leq 2\mathbb{E}\left\{(f(X) - m(X))^2\right\} + \mathbb{E}\{R(\mathcal{F}_n)\}$$
$$\leq 2\mathbb{E}\left\{(f(X) - m(X))^2\right\} + (16L + 4\sigma)\delta_n + \mathbb{E}\left\{R(\mathcal{F}_n')\right\}$$
$$\leq 2\mathbb{E}\left\{(f(X) - m(X))^2\right\} + (16L + 4\sigma)\delta_n + \mathbb{E}\left\{R_1(\mathcal{F}_n')\right\} + \mathbb{E}\left\{R_2(\mathcal{F}_n')\right\}.$$

Define, for all $f \in \mathcal{F}_n$ with $D(f) > D(m)$,

$$\tilde{\rho}(f) \triangleq \frac{\mathbb{E}\left\{(m(X) - f(X))^4\right\}}{\mathbb{E}\left\{(m(X) - f(X))^2\right\}}.$$

Since $|m(x)| \leq 1$ and $|f(x)| \leq 1$, we have that

$$\tilde{\rho}(f) \leq 4L^2.$$

Invoke the second part of Lemma 5.4 below to obtain

$$\mathbb{E}\left\{R_1(\mathcal{F}_n')\right\} \leq \max\left(8L^2, 4L^2 \sup_{f \in \mathcal{F}_n'} \tilde{\rho}(f)\right) \frac{\log M(\delta_n)}{n}$$
$$\leq \max\left(8L^2, 16L^2\right) \frac{\log M(\delta_n)}{n}$$
$$= 16L^2 \frac{\log M(\delta_n)}{n}.$$

By Condition (sG) and Lemma 5.3, we have for *all* $s > 0$,

$$\mathbb{E}\left\{\exp\left(s\varepsilon(f(X) - m(X))\right) | X\right\} \leq \exp(\lambda_0 s^2 (m(X) - f(X))^2 / 2).$$

For $|z| \leq 1$, apply the inequality $e^z \leq 1 + 2z$. Choose

$$s_0 = \frac{1}{L\sqrt{2\lambda_0}},$$

then

$$\frac{1}{2}\lambda_0 s^2 (f(X) - m(X))^2 \leq 1,$$

therefore, for $0 < s \leq s_0$,

$$\mathbb{E}\left\{\exp\left(s\varepsilon(f(X) - m(X))\right)\right\} \leq \mathbb{E}\left\{\exp\left(\frac{1}{2}\lambda_0 s^2 (f(X) - m(X))^2\right)\right\}$$
$$\leq 1 + \lambda_0 s^2 \mathbb{E}\left\{(f(X) - m(X))^2\right\}$$
$$\leq \exp\left(\lambda_0 s^2 \mathbb{E}\left\{(f(X) - m(X))^2\right\}\right).$$

Next, we invoke the first part of Lemma 5.4. We find that the value ϑ in Lemma 5.4 becomes

$$1/\vartheta = 8 \sup_{f \in \mathcal{F}'_n} \frac{\lambda_0 \mathbb{E}\{(f(X) - m(X))^2\}}{\mathbb{E}\{\Delta_f(X)\}} \leq 8\lambda_0,$$

and we obtain

$$\mathbb{E}\{R_2(\mathcal{F}'_n)\} \leq 4 \frac{\log M(\delta_n)}{n} \max\left(L\sqrt{2\lambda_0}, 8\lambda_0\right),$$

and this completes the proof of Proposition 5.4. $\square$

Instead of restricting the set of functions over which one minimizes, one can also add a penalty term to the functional to be minimized. Let $J_n(f) \geq 0$ be a penalty term penalizing the "roughness" of a function f. The *penalized modelling* or *penalized least-squares estimate*, m_n, is defined by

$$m_n = \arg\min_f \left\{\frac{1}{n} \sum_{i=1}^{n} |f(X_i) - Y_i|^2 + J_n(f)\right\}, \tag{5.16}$$

where one minimizes over all measurable functions f. Again, we do not require that the minimum in (5.16) be unique. In the case it is not unique, we randomly select one function which achieves the minimum.

A popular choice for $J_n(f)$ in the case $d = 1$ is

$$J_n(f) = \lambda_n \int |f''(t)|^2 dt, \tag{5.17}$$

where f'' denotes the second derivative of f and λ_n is some positive constant. One can show that for this penalty term the minimum in (5.16) is achieved by a cubic spline with knots at the X_i's, i.e., by a twice differentiable function which is equal to a polynomial of degree 3 (or less) between adjacent values of the X_i's (a so-called smoothing spline).

5.3. Universally Consistent Predictions: Bounded Y

5.3.1. *Partition-Based Prediction Strategies*

In this section we introduce our first prediction strategy for bounded ergodic processes. We assume throughout the section that $|Y_0|$ is bounded by a constant $B > 0$, with probability one, and that the bound B is known.

Generally, a prediction strategy is a sequence $g = \{g_t\}_{t=1}^{\infty}$ of functions

$$g_t : \left(\mathbb{R}^d\right)^t \times \mathbb{R}^{t-1} \to \mathbb{R}$$

so that the prediction formed at time t is $g_t(x_1^t, y_1^{t-1})$.

We assume that $(x_1, y_1), (x_2, y_2), \ldots$ are realizations of the random variables $(X_1, Y_1), (X_2, Y_2), \ldots$, such that $\{(X_n, Y_n)\}_{-\infty}^{\infty}$ is a jointly stationary and ergodic process.

After n time instants, the *normalized cumulative prediction error* is

$$L_n(g) = \frac{1}{n} \sum_{t=1}^{n} (g_t(X_1^t, Y_1^{t-1}) - Y_t)^2.$$

Our aim is to achieve small $L_n(g)$ when n is large.

Let us define our prediction strategy, at each time instant, as a convex combination of *elementary predictors*, where the weighting coefficients depend on the past performance of each elementary predictor.

We define an infinite array of elementary predictors $h^{(k,\ell)}$, $k, \ell = 1, 2, \ldots$ as follows. Let $\mathcal{P}_\ell = \{A_{\ell,j}, j = 1, 2, \ldots, m_\ell\}$ be a sequence of finite partitions of $\mathbb{R}$, and let $\mathcal{Q}_\ell = \{B_{\ell,j}, j = 1, 2, \ldots, m_\ell'\}$ be a sequence of finite partitions of $\mathbb{R}^d$. Introduce the corresponding quantizers:

$$F_\ell(y) = j, \text{ if } y \in A_{\ell,j}$$

and

$$G_\ell(x) = j, \text{ if } x \in B_{\ell,j} .$$

With some abuse of notation, for any n and $y_1^n \in \mathbb{R}^n$, we write $F_\ell(y_1^n)$ for the sequence $F_\ell(y_1), \ldots, F_\ell(y_n)$, and, similarly, for $x_1^n \in (\mathbb{R}^d)^n$, we write $G_\ell(x_1^n)$ for the sequence $G_\ell(x_1), \ldots, G_\ell(x_n)$.

Fix positive integers k, ℓ, and, for each $k+1$-long string z of positive integers and each k-long string s of positive integers, define the partitioning regression function estimate

$$\widehat{E}_n^{(k,\ell)}(x_1^n, y_1^{n-1}, z, s) = \frac{\sum_{\{k < t < n : G_\ell(x_{t-k}^t) = z, \, F_\ell(y_{t-k}^{t-1}) = s\}} y_t}{\left|\{k < t < n : G_\ell(x_{t-k}^t) = z, \, F_\ell(y_{t-k}^{t-1}) = s\}\right|},$$

for all $n > k+1$ where $0/0$ is defined to be 0.

Define the elementary predictor $h^{(k,\ell)}$ by

$$h_n^{(k,\ell)}(x_1^n, y_1^{n-1}) = \widehat{E}_n^{(k,\ell)}(x_1^n, y_1^{n-1}, G_\ell(x_{n-k}^n), F_\ell(y_{n-k}^{n-1})),$$

for $n = 1, 2, \ldots$. That is, $h_n^{(k,\ell)}$ quantizes the sequence x_1^n, y_1^{n-1} according to the partitions $\mathcal{Q}_\ell$ and $\mathcal{P}_\ell$, and looks for all appearances of the last seen quantized strings $G_\ell(x_{n-k}^n)$ of length $k+1$ and $F_\ell(y_{n-k}^{n-1})$ of length k in the past. Then it predicts according to the average of the y_t's following the string.

In contrast to the nonparametric regression estimation problem from i.i.d. data, for ergodic observations, it is impossible to choose $k = k_n$ and $\ell = \ell_n$ such that the corresponding predictor is universally-consistent for the class of bounded ergodic processes. The very important new principle is the combination or aggregation of elementary predictors (cf. [Cesa-Bianchi and Lugosi (2006)]). The proposed prediction algorithm proceeds as follows: let $\{q_{k,\ell}\}$ be a probability distribution on the set of all pairs (k, ℓ) of positive integers such that for all k, ℓ, $q_{k,\ell} > 0$. Put $c = 8B^2$, and define the weights

$$w_{t,k,\ell} = q_{k,\ell} e^{-(t-1)L_{t-1}(h^{(k,\ell)})/c} \tag{5.18}$$

and their normalized values

$$p_{t,k,\ell} = \frac{w_{t,k,\ell}}{W_t}, \tag{5.19}$$

where

$$W_t = \sum_{i,j=1}^{\infty} w_{t,i,j}. \tag{5.20}$$

The prediction strategy g is defined by

$$g_t(x_1^t, y_1^{t-1}) = \sum_{k,\ell=1}^{\infty} p_{t,k,\ell} h^{(k,\ell)}(x_1^t, y_1^{t-1}), \qquad t = 1, 2, \ldots \tag{5.21}$$

i.e., the prediction g_t is the convex linear combination of the elementary predictors such that an elementary predictor has non-negligible weight in the combination if it has good performance until time $t - 1$.

Theorem 5.1. [Györfi and Lugosi (2001)] *Assume that*

(a) the sequences of partition $\mathcal{P}_\ell$ is nested, that is, any cell of $\mathcal{P}_{\ell+1}$ is a subset of a cell of $\mathcal{P}_\ell$, $\ell = 1, 2, \ldots$;
(b) the sequences of partition $\mathcal{Q}_\ell$ is nested;
(c) the sequences of partition $\mathcal{P}_\ell$ is asymptotically fine;
(d) the sequences of partition $\mathcal{Q}_\ell$ is asymptotically fine;

Then the prediction scheme g defined above is universal with respect to the class of all jointly stationary and ergodic processes $\{(X_n, Y_n)\}_{-\infty}^{\infty}$ such that $|Y_0| \le B$.

One of the main ingredients of the proof is the following lemma, whose proof is a straightforward extension of standard arguments in the prediction theory of individual sequences; see, for example, [Kivinen and Warmuth (1999)], [Singer and Feder (2000)].

Lemma 5.5. *Let $\tilde{h}_1, \tilde{h}_2, \ldots$ be a sequence of prediction strategies (experts), and let $\{q_k\}$ be a probability distribution on the set of positive integers. Assume that $\tilde{h}_i(x_1^n, y_1^{n-1}) \in [-B, B]$ and $y_1^n \in [-B, B]^n$. Define*

$$w_{t,k} = q_k e^{-(t-1)L_{t-1}(\tilde{h}_k)/c}$$

with $c \geq 8B^2$, and

$$v_{t,k} = \frac{w_{t,k}}{\sum_{i=1}^{\infty} w_{t,i}}.$$

If the prediction strategy $\tilde{g}$ is defined by

$$\tilde{g}_t(x_1^n, y_1^{t-1}) = \sum_{k=1}^{\infty} v_{t,k} \tilde{h}_k(x_1^n, y_1^{t-1}) \qquad t = 1, 2, \ldots$$

then for every $n \geq 1$,

$$L_n(\tilde{g}) \leq \inf_k \left(L_n(\tilde{h}_k) - \frac{c \ln q_k}{n} \right).$$

Here $-\ln 0$ is treated as ∞.

Proof. Introduce $W_1 = 1$ and $W_t = \sum_{k=1}^{\infty} w_{t,k}$ for $t > 1$. First we show that for each $t > 1$,

$$\left[\sum_{k=1}^{\infty} v_{t,k} \left(y_t - \tilde{h}_k(x_1^n, y_1^{t-1}) \right) \right]^2 \leq -c \ln \frac{W_{t+1}}{W_t}. \qquad (5.22)$$

Note that

$$W_{t+1} = \sum_{k=1}^{\infty} w_{t,k} e^{-\left(y_t - \tilde{h}_k(x_1^n, y_1^{t-1})\right)^2/c} = W_t \sum_{k=1}^{\infty} v_{t,k} e^{-\left(y_t - \tilde{h}_k(x_1^n, y_1^{t-1})\right)^2/c},$$

so that

$$-c \ln \frac{W_{t+1}}{W_t} = -c \ln \left(\sum_{k=1}^{\infty} v_{t,k} e^{-\left(y_t - \tilde{h}_k(x_1^n, y_1^{t-1})\right)^2/c} \right).$$

Therefore, (5.22) becomes

$$\exp \left(\frac{-1}{c} \left[\sum_{k=1}^{\infty} v_{t,k} \left(y_t - \tilde{h}_k(x_1^n, y_1^{t-1}) \right) \right]^2 \right) \geq \sum_{k=1}^{\infty} v_{t,k} e^{-\left(y_t - \tilde{h}_k(x_1^n, y_1^{t-1})\right)^2/c},$$

which is implied by Jensen's inequality and the concavity of the function $F_t(z) = e^{-(y_t-z)^2/c}$ for $c \geq 8B^2$. Thus, (5.22) implies that

$$
\begin{aligned}
nL_n(\tilde{g}) &= \sum_{t=1}^{n} \left(y_t - \tilde{g}(x_1^n, y_1^{t-1}) \right)^2 \\
&= \sum_{t=1}^{n} \left[\sum_{k=1}^{\infty} v_{t,k} \left(y_t - \tilde{h}_k(x_1^n, y_1^{t-1}) \right) \right]^2 \\
&\leq -c \sum_{t=1}^{n} \ln \frac{W_{t+1}}{W_t} \\
&= -c \ln W_{n+1}
\end{aligned}
$$

and therefore

$$
\begin{aligned}
nL_n(\tilde{g}) &\leq -c \ln \left(\sum_{k=1}^{\infty} w_{n+1,k} \right) \\
&= -c \ln \left(\sum_{k=1}^{\infty} q_k e^{-nL_n(\tilde{h}_k)/c} \right) \\
&\leq -c \ln \left(\sup_k q_k e^{-nL_n(\tilde{h}_k)/c} \right) \\
&= \inf_k \left(-c \ln q_k + nL_n(\tilde{h}_k) \right) ,
\end{aligned}
$$

which concludes the proof.   □

Another main ingredient of the proof of Theorem 5.1 is known as Breiman's generalized ergodic theorem [Breiman (1957)], see also [Algoet (1994)] and [Györfi et al. (2002)].

Lemma 5.6. [Breiman (1957)] *Let* $Z = \{Z_i\}_{-\infty}^{\infty}$ *be a stationary and ergodic process. Let* T *denote the left shift operator. Let* f_i *be a sequence of real-valued functions such that for some function* f, $f_i(Z) \to f(Z)$ *almost surely. Assume that* $\mathbb{E}\{\sup_i |f_i(Z)|\} < \infty$. *Then*

$$
\lim_{t \to \infty} \frac{1}{n} \sum_{i=1}^{n} f_i(T^i Z) = \mathbb{E}\{f(Z)\} \qquad \text{almost surely.}
$$

Proof of Theorem 5.1. Because of (5.1), it is enough to show that

$$
\limsup_{n \to \infty} L_n(g) \leq L^* \qquad \text{almost surely.}
$$

By a double application of the ergodic theorem, as $n \to \infty$, almost surely,

$$
\widehat{E}_n^{(k,\ell)}(X_1^n, Y_1^{n-1}, z, s) = \frac{\frac{1}{n} \sum_{\{k<i<n: G_\ell(X_{t-k}^t)=z,\, F_\ell(Y_{t-k}^{t-1})=s\}} Y_i}{\frac{1}{n} \left| \{k < i < n : G_\ell(X_{t-k}^t) = z,\, F_\ell(Y_{t-k}^{t-1}) = s\} \right|}
$$

$$
\to \frac{E\{Y_0 I_{\{G_\ell(X_{-k}^0)=z,\, F_\ell(Y_{-k}^{-1})=s\}}\}}{\mathbb{P}\{G_\ell(X_{-k}^0) = z,\, F_\ell(Y_{-k}^{-1}) = s\}}
$$

$$
= E\{Y_0 | G_\ell(X_{-k}^0) = z,\, F_\ell(Y_{-k}^{-1}) = s\},
$$

and therefore

$$
\lim_{n\to\infty} \sup_z \sup_s |\widehat{E}_n^{(k,\ell)}(X_1^n, Y_1^{n-1}, z, s)
$$
$$
- E\{Y_0 | G_\ell(X_{-k}^0) = z,\, F_\ell(Y_{-k}^{-1}) = s\}| = 0
$$

almost surely. Thus, by Lemma 5.6, as $n \to \infty$, almost surely,

$$
L_n(h^{(k,\ell)}) = \frac{1}{n} \sum_{i=1}^n (h^{(k,\ell)}(X_1^i, Y_1^{i-1}) - Y_i)^2
$$

$$
= \frac{1}{n} \sum_{i=1}^n (\widehat{E}_n^{(k,\ell)}(X_1^i, Y_1^{i-1}, G_\ell(X_{i-k}^i), F_\ell(Y_{i-k}^{i-1})) - Y_i)^2
$$

$$
\to E\{(Y_0 - E\{Y_0 | G_\ell(X_{-k}^0),\, F_\ell(Y_{-k}^{-1})\})^2\}
$$

$$
\triangleq \epsilon_{k,\ell}.
$$

Since the partitions $\mathcal{P}_\ell$ and $\mathcal{Q}_\ell$ are nested, $E\{Y_0 | G_\ell(X_{-k}^0),\, F_\ell(Y_{-k}^{-1})\}$ is a martingale indexed by the pair (k, ℓ). Thus, the martingale convergence theorem (see, e.g., [Stout (1974)]) and assumption (c) and (d) for the sequence of partitions implies that

$$
\inf \epsilon_{k,\ell} = \lim_{k,\ell\to\infty} \epsilon_{k,\ell} = E\left\{\left(Y_0 - E\{Y_0 | X_{-\infty}^0,\, Y_{-\infty}^{-1}\}\right)^2\right\} = L^*.
$$

Now, by Lemma 5.5,

$$
L_n(g) \le \inf_{k,\ell} \left(L_n(h^{(k,\ell)}) - \frac{c \ln q_{k,\ell}}{n} \right), \tag{5.23}
$$

and therefore, almost surely,

$$
\begin{aligned}
\limsup_{n\to\infty} L_n(g) &\le \limsup_{n\to\infty} \inf_{k,\ell} \left(L_n(h^{(k,\ell)}) - \frac{c\ln q_{k,\ell}}{n} \right) \\
&\le \inf_{k,\ell} \limsup_{n\to\infty} \left(L_n(h^{(k,\ell)}) - \frac{c\ln q_{k,\ell}}{n} \right) \\
&\le \inf_{k,\ell} \limsup_{n\to\infty} L_n(h^{(k,\ell)}) \\
&= \inf_{k,\ell} \epsilon_{k,\ell} \\
&= \lim_{k,\ell\to\infty} \epsilon_{k,\ell} \\
&= L^*
\end{aligned}
$$

and the proof of the theorem is finished.   □

Theorem 5.1 shows that asymptotically, the predictor g_t defined by (5.21) predicts as well as the optimal predictor given by the regression function $\mathbb{E}\{Y_t|Y_{-\infty}^{t-1}\}$. In fact, g_t gives a good estimate of the regression function in the following (Cesáro) sense:

Corollary 5.1. *Under the conditions of Theorem 5.1*

$$
\lim_{n\to\infty} \frac{1}{n} \sum_{i=1}^{n} \left(\mathbb{E}\{Y_i|X_{-\infty}^i, Y_{-\infty}^{i-1}\} - g_i(X_1^i, Y_1^{i-1}) \right)^2 = 0 \qquad \text{almost surely.}
$$

Proof. By Theorem 5.1,

$$
\lim_{n\to\infty} \frac{1}{n} \sum_{i=1}^{n} \left(Y_i - g_i(X_1^i, Y_1^{i-1}) \right)^2 = L^* \qquad \text{almost surely.}
$$

Consider the following decomposition:

$$
\begin{aligned}
&\left(Y_i - g_i(X_1^i, Y_1^{i-1}) \right)^2 \\
&= \left(Y_i - \mathbb{E}\{Y_i|X_{-\infty}^i, Y_{-\infty}^{i-1}\} \right)^2 \\
&\quad + 2\left(Y_i - \mathbb{E}\{Y_i|X_{-\infty}^i, Y_{-\infty}^{i-1}\} \right) \left(\mathbb{E}\{Y_i|X_{-\infty}^i, Y_{-\infty}^{i-1}\} - g_i(X_1^i, Y_1^{i-1}) \right) \\
&\quad + \left(\mathbb{E}\{Y_i|X_{-\infty}^i, Y_{-\infty}^{i-1}\} - g_i(X_1^i, Y_1^{i-1}) \right)^2.
\end{aligned}
$$

Then the ergodic theorem implies that

$$
\lim_{n\to\infty} \frac{1}{n} \sum_{i=1}^{n} \left(Y_i - \mathbb{E}\{Y_i|X_{-\infty}^i, Y_{-\infty}^{i-1}\} \right)^2 = L^* \qquad \text{almost surely.}
$$

It remains to show that

$$\lim_{n\to\infty} \frac{1}{n} \sum_{i=1}^{n} \left(Y_i - \mathbb{E}\{Y_i|X_{-\infty}^i, Y_{-\infty}^{i-1}\}\right) \left(\mathbb{E}\{Y_i|Y_{-\infty}^{i-1}\} - g_i(X_1^i, Y_1^{i-1})\right) = 0.$$

(5.24)

almost surely. However this is a straightforward consequence of Kolmogorov's classical strong law of large numbers for martingale differences due to [Chow (1965)] (see also Theorem 3.3.1 in [Stout (1974)]). It states that, if $\{Z_i\}$ is, a martingale difference sequence with

$$\sum_{n=1}^{\infty} \frac{\mathbb{E}Z_n^2}{n^2} < \infty,$$

(5.25)

then

$$\lim_{n\to\infty} \frac{1}{n} \sum_{i=1}^{n} Z_i = 0 \qquad \text{almost surely.}$$

Thus, (5.24) is implied by Chow's theorem since the martingale differences $Z_i = \left(Y_i - \mathbb{E}\{Y_i|X_{-\infty}^i, Y_{-\infty}^{i-1}\}\right) \left(\mathbb{E}\{Y_i|X_{-\infty}^i, Y_{-\infty}^{i-1}\} - g_i(X_1^i, Y_1^{i-1})\right)$ are bounded by $4B^2$. (To see that the Z_i's indeed form a martingale difference sequence just note that $\mathbb{E}\{Z_i|X_{-\infty}^i, Y_{-\infty}^{i-1}\} = 0$ for all i.) $\qquad\square$

Remark 5.1. (Choice of $q_{k,\ell}$) Theorem 5.1 is true independently of the choice of the $q_{k,\ell}$'s as long as these values are strictly positive for all k and ℓ. In practice, however, the choice of $q_{k,\ell}$ may have an impact on the performance of the predictor. For example, if the distribution $\{q_{k,\ell}\}$ has a very rapidly decreasing tail, then the term $-\ln q_{k,\ell}/n$ will be large for moderately large values of k and ℓ, and the performance of g will be determined by the best of just a few of the elementary predictors $h^{(k,\ell)}$. Thus, it may be advantageous to choose $\{q_{k,\ell}\}$ to be a large-tailed distribution. For example, $q_{k,\ell} = c_0 k^{-2} \ell^{-2}$ is a safe choice, where c_0 is an appropriate normalizing constant.

5.3.2. *Kernel-Based Prediction Strategies*

We introduce in this section a class of *kernel-based* prediction strategies for stationary and ergodic sequences. The main advantage of this approach in contrast to the partition-based strategy is that it replaces the rigid discretization of the past appearances by more flexible rules. This also often leads to faster algorithms in practical applications.

To simplify the notation, we start with the simple "moving-window" scheme, corresponding to a uniform kernel function, and treat the general case briefly later. As before, we define an array of experts $h^{(k,\ell)}$, where k and ℓ are positive integers. We associate to each pair (k,ℓ) two radii $r_{k,\ell} > 0$ and $r'_{k,\ell} > 0$ such that, for any fixed k,

$$\lim_{\ell \to \infty} r_{k,\ell} = 0, \qquad (5.26)$$

and

$$\lim_{\ell \to \infty} r'_{k,\ell} = 0. \qquad (5.27)$$

Finally, let the location of the matches be

$$J_n^{(k,\ell)} = \left\{ t : k < t < n, \|x_{t-k}^t - x_{n-k}^n\| \leq r_{k,\ell}, \ \|y_{t-k}^{t-1} - y_{n-k}^{n-1}\| \leq r'_{k,\ell} \right\} .$$

Then the elementary expert $h_n^{(k,\ell)}$ at time n is defined by

$$h_n^{(k,\ell)}(x_1^n, y_1^{n-1}) = \frac{\sum_{\{t \in J_n^{(k,\ell)}\}} y_t}{|J_n^{(k,\ell)}|}, \qquad n > k + 1, \qquad (5.28)$$

where $0/0$ is defined to be 0. The pool of experts is mixed the same way as in the case of the partition-based strategy (cf. (5.18), (5.19), (5.20) and (5.21)).

Theorem 5.2. *Suppose that (5.26) and (5.27) are verified. Then the kernel-based strategy defined above is universally-consistent with respect to the class of all jointly stationary and ergodic processes $\{(X_n, Y_n)\}_{-\infty}^{\infty}$ such that $|Y_0| \leq B$.*

Remark 5.2. This theorem may be extended to a more general class of kernel-based strategies. Define a *kernel function* as any map $K : \mathbb{R}_+ \to \mathbb{R}_+$. The kernel-based strategy parallels the moving-window scheme defined above, with the only difference that, in definition (5.28) of the elementary strategy, the regression function estimate is replaced by

$$h_n^{(k,\ell)}(x_1^n, y_1^{n-1})$$

$$= \frac{\sum_{\{k < t < n\}} K\left(\|x_{t-k}^t - x_{n-k}^n\|/r_{k,\ell}\right) K\left(\|y_{t-k}^{t-1} - y_{n-k}^{n-1}\|/r'_{k,\ell}\right) y_t}{\sum_{\{k < t < n\}} K\left(\|x_{t-k}^t - x_{n-k}^n\|/r_{k,\ell}\right) K\left(\|y_{t-k}^{t-1} - y_{n-k}^{n-1}\|/r'_{k,\ell}\right)} .$$

Observe that, if K is the naïve kernel $K(x) = I_{\{x \leq 1\}}$, we recover the moving-window strategy discussed above. Typical nonuniform kernels assign a smaller weight to the observations x_{t-k}^t and y_{t-k}^{t-1} whose distance from x_{n-k}^n and y_{n-k}^{n-1} is larger. Such kernels promise a better prediction of the local structure of the conditional distribution.

5.3.3. *Nearest-Neighbor-Based Prediction Strategy*

This strategy is yet more robust with respect to the kernel strategy and thus, also with respect to the partition strategy, since it does not suffer from the scaling problem as partition and kernel-based strategies where the quantizer and the radius have to be carefully chosen to obtain "good" performance. As well as this, in practical applications it runs extremely fast compared with the kernel and partition schemes as it is much less likely to be slowed down by calculations for certain experts.

To introduce the strategy, we start again by defining an infinite array of experts $h^{(k,\ell)}$, where k and ℓ are positive integers. As before, k is the length of the past observation vectors being scanned by the elementary expert and, for each ℓ, choose $p_\ell \in (0,1)$ such that

$$\lim_{\ell \to \infty} p_\ell = 0 \,, \tag{5.29}$$

and set

$$\bar{\ell} = \lfloor p_\ell n \rfloor$$

(where $\lfloor . \rfloor$ is the floor function). At time n, for fixed k and ℓ ($n > k + \bar{\ell} + 1$), the expert searches for the $\bar{\ell}$ nearest-neighbors (NN) of the last seen observation x_{n-k}^n and y_{n-k}^{n-1} in the past and predicts accordingly. More precisely, let

$$J_n^{(k,\ell)} = \{\, t : k < t < n, (x_{t-k}^t, y_{t-k}^{t-1}) \text{ is among the } \bar{\ell} \text{ NN of } (x_{n-k}^n, y_{n-k}^{n-1}) \text{ in}$$
$$(x_1^{k+1}, y_1^k), \dots, (x_{n-k-1}^{n-1}, y_{n-k-1}^{n-2})\}$$

and introduce the elementary predictor

$$h_n^{(k,\ell)}(x_1^n, y_1^{n-1}) = \frac{\sum_{\{t \in J_n^{(k,\ell)}\}} y_t}{|J_n^{(k,\ell)}|}$$

if the sum is nonvoid, and 0 otherwise. Finally, the experts are mixed as before (cf. (5.18), (5.19), (5.20) and (5.21)).

Theorem 5.3. *Suppose that (5.29) is verified and that for each vector* **s** *the random variable*

$$\|(X_1^{k+1}, Y_1^k) - \mathbf{s}\|$$

has a continuous distribution function. Then, the nearest-neighbor strategy defined above is universally-consistent with respect to the class of all jointly stationary and ergodic processes $\{(X_n, Y_n)\}_{-\infty}^\infty$ such that $|Y_0| \le B$.

5.3.4. Generalized Linear Estimates

This section is devoted to an alternative way of defining a universal predictor for stationary and ergodic processes. It is in effect an extension of the approach presented in [Győrfi and Lugosi (2001)]. Once again, we apply the method described in the previous sections to combine elementary predictors, but now we use elementary predictors which are generalized linear predictors. More precisely, we define an infinite array of elementary experts $h^{(k,\ell)}$, $k, \ell = 1, 2, \ldots$ as follows. Let $\{\phi_j^{(k)}\}_{j=1}^{\ell}$ be real-valued functions defined on $(\mathbb{R}^d)^{(k+1)} \times \mathbb{R}^k$. The elementary predictor $h_n^{(k,\ell)}$ generates a prediction of form

$$h_n^{(k,\ell)}(x_1^n, y_1^{n-1}) = \sum_{j=1}^{\ell} c_{n,j} \phi_j^{(k)}(x_{n-k}^n, y_{n-k}^{n-1}),$$

where the coefficients $c_{n,j}$ are calculated according to the past observations x_1^n, y_1^{n-1}. More precisely, the coefficients $c_{n,j}$ are defined as the real numbers which minimize the criterion

$$\sum_{t=k+1}^{n-1} \left(\sum_{j=1}^{\ell} c_j \phi_j^{(k)}(x_{t-k}^t, y_{t-k}^{t-1}) - y_t \right)^2 \tag{5.30}$$

if $n > k + 1$, and the all-zero vector otherwise. It can be shown using a recursive technique (see, e.g., [Tsypkin (1971)], [Győrfi (1984)], [Singer and Feder (2000)], and [Győrfi and Lugosi (2001)]) that the $c_{n,j}$ can be calculated with small computational complexity.

The experts are mixed via an exponential weighting, which is defined the same way as earlier (cf. (5.18), (5.19), (5.20) and (5.21)).

Theorem 5.4. [Győrfi and Lugosi (2001)] *Suppose that $|\phi_j^{(k)}| \leq 1$ and, for any fixed k, suppose that the set*

$$\left\{ \sum_{j=1}^{\ell} c_j \phi_j^{(k)}; \quad (c_1, \ldots, c_\ell), \ \ell = 1, 2, \ldots \right\}$$

is dense in the set of continuous functions of $d(k + 1) + k$ variables. Then the generalized linear strategy defined above is universally-consistent with respect to the class of all jointly stationary and ergodic processes $\{(X_n, Y_n)\}_{-\infty}^{\infty}$ such that $|Y_0| \leq B$.

5.4. Universally Consistent Predictions: Unbounded Y

5.4.1. *Partition-Based Prediction Strategies*

Let $\widehat{E}_n^{(k,\ell)}(x_1^n, y_1^{n-1}, z, s)$ be defined as in Section 5.3.1. Introduce the truncation function

$$
T_m(z) = \begin{cases} m & \text{if } z > m \\ z & \text{if } |z| < m \\ -m & \text{if } z < -m, \end{cases}
$$

Define the elementary predictor $h^{(k,\ell)}$ by

$$
h_n^{(k,\ell)}(x_1^n, y_1^{n-1}) = T_{n^\delta}\left(\widehat{E}_n^{(k,\ell)}(x_1^n, y_1^{n-1}, G_\ell(x_{n-k}^n), F_\ell(y_{n-k}^{n-1}))\right),
$$

where

$$
0 < \delta < 1/8,
$$

for $n = 1, 2, \ldots$. That is, $h_n^{(k,\ell)}$ is the truncation of the elementary predictor introduced in Section 5.3.1.

The proposed prediction algorithm proceeds as follows: let $\{q_{k,\ell}\}$ be a probability distribution on the set of all pairs (k, ℓ) of positive integers such that for all k, ℓ, $q_{k,\ell} > 0$. For a time-dependent learning parameter $\eta_t > 0$, define the weights

$$
w_{t,k,\ell} = q_{k,\ell} e^{-\eta_t (t-1) L_{t-1}(h^{(k,\ell)})} \tag{5.31}
$$

and their normalized values

$$
p_{t,k,\ell} = \frac{w_{t,k,\ell}}{W_t}, \tag{5.32}
$$

where

$$
W_t = \sum_{i,j=1}^{\infty} w_{t,i,j}. \tag{5.33}
$$

One can change the speed of the learning with the tuning of the learning parameter η_t. Due to "large" η_t the learning process is "faster", but it may sacrifice the convergence. In the case of "small" η_t the learning process is slower, but it converges. In theoretical viewpoint we would like to find the greatest η_t that still satisfies the convergence criteria.

The prediction strategy g is defined by

$$
g_t(x_1^t, y_1^{t-1}) = \sum_{k,\ell=1}^{\infty} p_{t,k,\ell} h^{(k,\ell)}(x_1^t, y_1^{t-1}), \qquad t = 1, 2, \ldots \tag{5.34}
$$

Theorem 5.5. [Győrfi and Ottucsák (2007)] *Assume that the conditions (a), (b), (c) and (d) of Theorem 5.1 are satisfied. Choose $\eta_t = 1/\sqrt{t}$. Then the prediction scheme g defined above is universally-consistent with respect to the class of all ergodic processes $\{(X_n, Y_n)\}_{-\infty}^{\infty}$ such that*

$$\mathbb{E}\{Y_1^4\} < \infty.$$

Here we describe a result, which is used in the analysis. This lemma is a modification of the analysis of [Auer *et al.* (2002)], which allows of handling the case when the learning parameter of the algorithm (η_t) is time-dependent and the number of the elementary predictors is infinite.

Lemma 5.7. [Győrfi and Ottucsák (2007)] *Let $h^{(1)}, h^{(2)}, \ldots$ be a sequence of prediction strategies (experts). Let $\{q_k\}$ be a probability distribution on the set of positive integers. Denote the normalized loss of the expert $h = (h_1, h_2, \ldots)$ by*

$$L_n(h) = \frac{1}{n} \sum_{t=1}^{n} \ell_t(h),$$

where

$$\ell_t(h) = \ell(h_t, Y_t)$$

and the loss function ℓ is convex in its first argument h. Define

$$w_{t,k} = q_k e^{-\eta_t(t-1)L_{t-1}(h^{(k)})}$$

where $\eta_t > 0$ is monotonically decreasing, and

$$p_{t,k} = \frac{w_{t,k}}{W_t}$$

where

$$W_t = \sum_{k=1}^{\infty} w_{t,k} \ .$$

If the prediction strategy $g = (g_1, g_2, \ldots)$ is defined by

$$g_t = \sum_{k=1}^{\infty} p_{t,k} h_t^{(k)} \qquad t = 1, 2, \ldots$$

then, for every $n \geq 1$,

$$L_n(g) \leq \inf_k \left(L_n(h^{(k)}) - \frac{\ln q_k}{n\eta_{n+1}} \right) + \frac{1}{2n} \sum_{t=1}^{n} \eta_t \sum_{k=1}^{\infty} p_{t,k} \ell_t^2(h^{(k)}).$$

Proof. Introduce some notations:

$$w'_{t,k} = q_k e^{-\eta_{t-1}(t-1)L_{t-1}(h^{(k)})},$$

which is the weight $w_{t,k}$, where η_t is replaced by η_{t-1} and the sum of these are

$$W'_t = \sum_{k=1}^{\infty} w'_{t,k}.$$

We start the proof with the following chain of bounds:

$$\frac{1}{\eta_t} \ln \frac{W'_{t+1}}{W_t} = \frac{1}{\eta_t} \ln \frac{\sum_{k=1}^{\infty} w_{t,k} e^{-\eta_t \ell_t(h^{(k)})}}{W_t}$$

$$= \frac{1}{\eta_t} \ln \sum_{k=1}^{\infty} p_{t,k} e^{-\eta_t \ell_t(h^{(k)})}$$

$$\leq \frac{1}{\eta_t} \ln \sum_{k=1}^{\infty} p_{t,k} \left(1 - \eta_t \ell_t(h^{(k)}) + \frac{\eta_t^2}{2} \ell_t^2(h^{(k)}) \right)$$

because of $e^{-x} \leq 1 - x + x^2/2$ for $x \geq 0$. Moreover,

$$\frac{1}{\eta_t} \ln \frac{W'_{t+1}}{W_t}$$

$$\leq \frac{1}{\eta_t} \ln \left(1 - \eta_t \sum_{k=1}^{\infty} p_{t,k} \ell_t(h^{(k)}) + \frac{\eta_t^2}{2} \sum_{k=1}^{\infty} p_{t,k} \ell_t^2(h^{(k)}) \right)$$

$$\leq - \sum_{k=1}^{\infty} p_{t,k} \ell_t(h^{(k)}) + \frac{\eta_t}{2} \sum_{k=1}^{\infty} p_{t,k} \ell_t^2(h^{(k)}) \tag{5.35}$$

$$= - \sum_{k=1}^{\infty} p_{t,k} \ell(h_t^{(k)}, Y_t) + \frac{\eta_t}{2} \sum_{k=1}^{\infty} p_{t,k} \ell_t^2(h^{(k)})$$

$$\leq - \ell \left(\sum_{k=1}^{\infty} p_{t,k} h_t^{(k)}, Y_t \right) + \frac{\eta_t}{2} \sum_{k=1}^{\infty} p_{t,k} \ell_t^2(h^{(k)}) \tag{5.36}$$

$$= - \ell_t(g) + \frac{\eta_t}{2} \sum_{k=1}^{\infty} p_{t,k} \ell_t^2(h^{(k)}) \tag{5.37}$$

where (5.35) follows from the fact that $\ln(1 + x) \leq x$ for all $x > -1$ and in (5.36) we used the convexity of the loss $\ell(h, y)$ in its first argument h. From (5.37) after rearranging we obtain

$$\ell_t(g) \leq - \frac{1}{\eta_t} \ln \frac{W'_{t+1}}{W_t} + \frac{\eta_t}{2} \sum_{k=1}^{\infty} p_{t,k} \ell_t^2(h^{(k)}) \ .$$

Then write a telescope formula:

$$\frac{1}{\eta_t} \ln W_t - \frac{1}{\eta_t} \ln W'_{t+1} = \left(\frac{1}{\eta_t} \ln W_t - \frac{1}{\eta_{t+1}} \ln W_{t+1} \right)$$
$$+ \left(\frac{1}{\eta_{t+1}} \ln W_{t+1} - \frac{1}{\eta_t} \ln W'_{t+1} \right)$$
$$= (A_t) + (B_t).$$

We have that

$$\sum_{t=1}^{n} A_t = \sum_{t=1}^{n} \left(\frac{1}{\eta_t} \ln W_t - \frac{1}{\eta_{t+1}} \ln W_{t+1} \right)$$
$$= \frac{1}{\eta_1} \ln W_1 - \frac{1}{\eta_{n+1}} \ln W_{n+1}$$
$$= -\frac{1}{\eta_{n+1}} \ln \sum_{k=1}^{\infty} q_k e^{-\eta_{n+1} n L_n(h^{(k)})}$$
$$\leq -\frac{1}{\eta_{n+1}} \ln \sup_k q_k e^{-\eta_{n+1} n L_n(h^{(k)})}$$
$$= -\frac{1}{\eta_{n+1}} \sup_k \left(\ln q_k - \eta_{n+1} n L_n(h^{(k)}) \right)$$
$$= \inf_k \left(n L_n(h^{(k)}) - \frac{\ln q_k}{\eta_{n+1}} \right).$$

$\frac{\eta_{t+1}}{\eta_t} \leq 1$, therefore applying Jensen's inequality for a concave function, we obtain

$$W_{t+1} = \sum_{i=1}^{\infty} q_i e^{-\eta_{t+1} t L_t(h^{(i)})}$$
$$= \sum_{i=1}^{\infty} q_i \left(e^{-\eta_t t L_t(h^{(i)})} \right)^{\frac{\eta_{t+1}}{\eta_t}}$$
$$\leq \left(\sum_{i=1}^{\infty} q_i e^{-\eta_t t L_t(h^{(i)})} \right)^{\frac{\eta_{t+1}}{\eta_t}}$$
$$= \left(W'_{t+1} \right)^{\frac{\eta_{t+1}}{\eta_t}}.$$

Thus,

$$B_t = \frac{1}{\eta_{t+1}} \ln W_{t+1} - \frac{1}{\eta_t} \ln W'_{t+1}$$

$$\leq \frac{1}{\eta_{t+1}} \frac{\eta_{t+1}}{\eta_t} \ln W'_{t+1} - \frac{1}{\eta_t} \ln W'_{t+1}$$

$$= 0.$$

We can summarize the bounds:

$$L_n(g) \leq \inf_k \left(L_n(h^{(k)}) - \frac{\ln q_k}{n \eta_{n+1}} \right) + \frac{1}{2n} \sum_{t=1}^{n} \eta_t \sum_{k=1}^{\infty} p_{t,k} \ell_t^2(h^{(k)}) .$$

$\square$

Proof of Theorem 5.5. Because of (5.1), it is enough to show that

$$\limsup_{n \to \infty} L_n(g) \leq L^* \qquad \text{almost surely.}$$

Because of the proof of Theorem 5.1, as $n \to \infty$, almost surely,

$$\widehat{E}_n^{(k,\ell)}(X_1^n, Y_1^{n-1}, z, s) \to \mathbb{E}\{Y_0 \mid G_\ell(X_{-k}^0) = z, F_\ell(Y_{-k}^{-1}) = s\},$$

and therefore for all z and s

$$T_{n^\delta}\left(\widehat{E}_n^{(k,\ell)}(X_1^n, Y_1^{n-1}, z, s) \right) \to \mathbb{E}\{Y_0 \mid G_\ell(X_{-k}^0) = z, F_\ell(Y_{-k}^{-1}) = s\}.$$

By Lemma 5.6, as $n \to \infty$, almost surely,

$$L_n(h^{(k,\ell)})$$

$$= \frac{1}{n} \sum_{t=1}^{n} (h^{(k,\ell)}(X_1^t, Y_1^{t-1}) - Y_t)^2$$

$$= \frac{1}{n} \sum_{t=1}^{n} \left(T_{t^\delta}\left(\widehat{E}_t^{(k,\ell)}(X_1^t, Y_1^{t-1}, G_\ell(X_{t-k}^t), F_\ell(Y_{t-k}^{t-1})) \right) - Y_t \right)^2$$

$$\to \mathbb{E}\{(Y_0 - \mathbb{E}\{Y_0 \mid G_\ell(X_{-k}^0), F_\ell(Y_{-k}^{-1})\})^2\}$$

$$\triangleq \epsilon_{k,\ell}.$$

In the same way as in the proof of Theorem 5.1, we obtain

$$\inf_{k,l} \epsilon_{k,l} = \lim_{k,\ell \to \infty} \epsilon_{k,\ell} = \mathbb{E}\left\{ (Y_0 - \mathbb{E}\{Y_0 \mid X_{-\infty}^0, Y_{-\infty}^{-1}\})^2 \right\} = L^*.$$

Apply Lemma 5.7 with $\eta_t = \frac{1}{\sqrt{t}}$ and for the squared loss $\ell_t(h) = (h_t - Y_t)^2$, then the square loss is convex in its first argument h, so

$$L_n(g) \leq \inf_{k,\ell} \left(L_n(h^{(k,\ell)}) - \frac{2 \ln q_{k,\ell}}{\sqrt{n}} \right)$$

$$+ \frac{1}{2n} \sum_{t=1}^{n} \frac{1}{\sqrt{t}} \sum_{k,\ell=1}^{\infty} p_{t,k,\ell}\big(h^{(k,\ell)}(X_1^t, Y_1^{t-1}) - Y_t\big)^4 . \quad (5.38)$$

On the one hand, almost surely,

$$\limsup_{n \to \infty} \inf_{k,\ell} \left(L_n(h^{(k,\ell)}) - \frac{2 \ln q_{k,\ell}}{\sqrt{n}} \right)$$

$$\leq \inf_{k,\ell} \limsup_{n \to \infty} \left(L_n(h^{(k,\ell)}) - \frac{2 \ln q_{k,\ell}}{\sqrt{n}} \right)$$

$$= \inf_{k,\ell} \limsup_{n \to \infty} L_n(h^{(k,\ell)})$$

$$= \inf_{k,\ell} \epsilon_{k,\ell}$$

$$= \lim_{k,\ell \to \infty} \epsilon_{k,\ell}$$

$$= L^* .$$

On the other hand,

$$\frac{1}{n} \sum_{t=1}^{n} \frac{1}{\sqrt{t}} \sum_{k,\ell} p_{t,k,\ell}(h^{(k,\ell)}(X_1^t, Y_1^{t-1}) - Y_t)^4$$

$$\leq \frac{8}{n} \sum_{t=1}^{n} \frac{1}{\sqrt{t}} \sum_{k,\ell} p_{t,k,\ell} \left(h^{(k,\ell)}(X_1^t, Y_1^{t-1})^4 + Y_t^4 \right)$$

$$\leq \frac{8}{n} \sum_{t=1}^{n} \frac{1}{\sqrt{t}} \sum_{k,\ell} p_{t,k,\ell} \left(t^{4\delta} + Y_t^4 \right)$$

$$= \frac{8}{n} \sum_{t=1}^{n} \frac{t^{4\delta} + Y_t^4}{\sqrt{t}} ,$$

therefore, almost surely,

$$\limsup_{n \to \infty} \frac{1}{n} \sum_{t=1}^{n} \frac{1}{\sqrt{t}} \sum_{k,\ell} p_{t,k,\ell}(h^{(k,\ell)}(X_1^t, Y_1^{t-1}) - Y_t)^4$$

$$\leq \limsup_{n \to \infty} \frac{8}{n} \sum_{t=1}^{n} \frac{Y_t^4}{\sqrt{t}}$$

$$= 0,$$

where we applied that $\mathbb{E}\{Y_1^4\} < \infty$ and $0 < \delta < \frac{1}{8}$. Summarizing these bounds, we obtain, almost surely,

$$\limsup_{n\to\infty} L_n(g) \leq L^*$$

and the proof of the theorem is finished. $\qquad\square$

Corollary 5.2. [Györfi and Ottucsák (2007)] *Under the conditions of Theorem 5.5,*

$$\lim_{n\to\infty} \frac{1}{n} \sum_{t=1}^{n} \left(\mathbb{E}\{Y_t \mid X_{-\infty}^t, Y_{-\infty}^{t-1}\} - g_t(X_1^t, Y_1^{t-1})\right)^2 = 0 \qquad almost \ surely.$$

$$(5.39)$$

Proof. By Theorem 5.5,

$$\lim_{n\to\infty} \frac{1}{n} \sum_{t=1}^{n} \left(Y_t - g_t(X_1^t, Y_1^{t-1})\right)^2 = L^* \qquad almost \ surely \qquad (5.40)$$

and by the ergodic theorem we have

$$\lim_{n\to\infty} \frac{1}{n} \sum_{t=1}^{n} \mathbb{E}\left\{ \left(Y_t - \mathbb{E}\{Y_t \mid X_{-\infty}^t, Y_{-\infty}^{t-1}\}\right)^2 \mid X_{-\infty}^t, Y_{-\infty}^{t-1} \right\} = L^* \qquad (5.41)$$

almost surely. Now we may write as $n \to \infty$, that

$$\frac{1}{n} \sum_{t=1}^{n} \left(\mathbb{E}\{Y_t \mid X_{-\infty}^t, Y_{-\infty}^{t-1}\} - g_t(X_1^t, Y_1^{t-1})\right)^2$$

$$= \frac{1}{n} \sum_{t=1}^{n} \mathbb{E}\{\left(Y_t - g_t(X_1^t, Y_1^{t-1})\right)^2 \mid X_{-\infty}^t, Y_{-\infty}^{t-1}\}$$

$$\quad - \frac{1}{n} \sum_{t=1}^{n} \mathbb{E}\{\left(Y_t - \mathbb{E}\{Y_t \mid X_{-\infty}^t, Y_{-\infty}^{t-1}\}\right)^2 \mid X_{-\infty}^t, Y_{-\infty}^{t-1}\}$$

$$= \frac{1}{n} \sum_{t=1}^{n} \mathbb{E}\{\left(Y_t - g_t(X_1^t, Y_1^{t-1})\right)^2 \mid X_{-\infty}^t, Y_{-\infty}^{t-1}\}$$

$$\quad - \frac{1}{n} \sum_{t=1}^{n} \left(Y_t - g_t(X_1^t, Y_1^{t-1})\right)^2 + o(1) \qquad (5.42)$$

$$= 2\frac{1}{n} \sum_{t=1}^{n} g_t(X_1^t, Y_1^{t-1})(Y_t - \mathbb{E}\{Y_t \mid X_{-\infty}^t, Y_{-\infty}^{t-1}\})$$

$$\quad - \frac{1}{n} \sum_{t=1}^{n} \left(Y_t^2 - \mathbb{E}\{Y_t^2 \mid X_{-\infty}^t, Y_{-\infty}^{t-1}\}\right) + o(1) \qquad almost \ surely$$

where (5.42) holds because of (5.40) and (5.41). The second sum is

$$\frac{1}{n} \sum_{t=1}^{n} \left(Y_t^2 - \mathbb{E}\{Y_t^2 \mid X_{-\infty}^t, Y_{-\infty}^{t-1}\} \right) \to 0 \qquad \text{almost surely}$$

by the ergodic theorem. Put

$$Z_t = g_t(X_1^t, Y_1^{t-1})(Y_t - \mathbb{E}\{Y_t \mid X_{-\infty}^t, Y_{-\infty}^{t-1}\}).$$

In order to finish the proof it suffices to show

$$\lim_{n \to \infty} \frac{1}{n} \sum_{t=1}^{n} Z_t = 0 . \tag{5.43}$$

Then

$$\mathbb{E}\{Z_t \mid X_{-\infty}^t, Y_{-\infty}^{t-1}\} = 0,$$

for all t, so the Z_t's form a martingale difference sequence. By the strong law of large numbers for martingale differences due to [Chow (1965)], one has to verify (5.25). By the construction of g_n,

$$\mathbb{E}\left\{ Z_n^2 \right\} = \mathbb{E}\left\{ \left(g_n(X_1^n, Y_1^{n-1})(Y_n - \mathbb{E}\{Y_n \mid X_{-\infty}^n, Y_{-\infty}^{n-1}\}) \right)^2 \right\}$$
$$\leq \mathbb{E}\left\{ g_n(X_1^n, Y_1^{n-1})^2 Y_n^2 \right\}$$
$$\leq n^{2\delta} \mathbb{E}\left\{ Y_1^2 \right\},$$

therefore (5.25) is verified, (5.43) is proved and the proof of the corollary is finished.   □

5.4.2. *Kernel-Based Prediction Strategies*

Apply the notations of Section 5.3.2. Then, the elementary expert $h_n^{(k,\ell)}$ at time n is defined by

$$h_n^{(k,\ell)}(x_1^n, y_1^{n-1}) = T_{\min\{n^\delta, \ell\}} \left(\frac{\sum_{\{t \in J_n^{(k,\ell)}\}} y_t}{|J_n^{(k,\ell)}|} \right), \qquad n > k + 1,$$

where $0/0$ is defined to be 0 and $0 < \delta < 1/8$. The pool of experts is mixed the same way as in the case of the partition-based strategy (cf. (5.31), (5.32), (5.33) and (5.34)).

Theorem 5.6. [Biau *et al.* (2010)] *Choose $\eta_t = 1/\sqrt{t}$ and suppose that (5.26) and (5.27) are verified. Then, the kernel-based strategy defined above is universally-consistent with respect to the class of all jointly stationary and ergodic processes $\{(X_n, Y_n)\}_{-\infty}^{\infty}$ such that*

$$\mathbb{E}\{Y_0^4\} < \infty.$$

5.4.3. *Nearest-Neighbor-Based Prediction Strategy*

Apply the notations of Section 5.3.3. Then the elementary expert $h_n^{(k,\ell)}$ at time n is defined by

$$h_n^{(k,\ell)}(x_1^n, y_1^{n-1}) = T_{\min\{n^\delta, \ell\}}\left(\frac{\sum_{\{t \in J_n^{(k,\ell)}\}} y_t}{|J_n^{(k,\ell)}|}\right), \qquad n > k+1,$$

if the sum is nonvoid, and 0 otherwise and $0 < \delta < 1/8$. The pool of experts is mixed the same way as in the case of the histogram-based strategy (cf. (5.31), (5.32), (5.33) and (5.34)).

Theorem 5.7. [Biau *et al.* (2010)] *Choose $\eta_t = 1/\sqrt{t}$ and suppose that (5.29) is verified. Suppose also that, for each vector* **s**, *the random variable*

$$\|(X_1^{k+1}, Y_1^k) - \mathbf{s}\|$$

has a continuous distribution function. Then, the nearest-neighbor strategy defined above is universally-consistent with respect to the class of all jointly stationary and ergodic processes $\{(X_n, Y_n)\}_{-\infty}^\infty$ such that

$$\mathbb{E}\{Y_0^4\} < \infty.$$

5.4.4. *Generalized Linear Estimates*

Apply the notations of Section 5.3.4. The elementary predictor $h_n^{(k,\ell)}$ generates a prediction of form

$$h_n^{(k,\ell)}(x_1^n, y_1^{n-1}) = T_{\min\{n^\delta, \ell\}}\left(\sum_{j=1}^\ell c_{n,j} \phi_j^{(k)}(x_{n-k}^n, y_{n-k}^{n-1})\right),$$

with $0 < \delta < 1/8$. The pool of experts is mixed the same way as in the case of the histogram-based strategy (cf. (5.31), (5.32), (5.33) and (5.34)).

Theorem 5.8. [Biau *et al.* (2010)] *Choose $\eta_t = 1/\sqrt{t}$ and suppose that $|\phi_j^{(k)}| \le 1$ and, for any fixed k, suppose that the set*

$$\left\{\sum_{j=1}^\ell c_j \phi_j^{(k)}; \ (c_1, \ldots, c_\ell), \ \ell = 1, 2, \ldots\right\}$$

is dense in the set of continuous functions of $d(k+1) + k$ variables. Then, the generalized linear strategy defined above is universally-consistent with respect to the class of all jointly stationary and ergodic processes $\{(X_n, Y_n)\}_{-\infty}^\infty$ such that

$$\mathbb{E}\{Y_0^4\} < \infty.$$

5.4.5. *Prediction of Gaussian Processes*

We consider in this section the classical problem of Gaussian time series prediction. In this context, parametric models based on distribution assumptions and structural conditions such as AR(p), MA(q), ARMA(p,q) and ARIMA(p,d,q) are usually fitted to the data (cf. [Gerencsér and Rissanen (1986)], [Gerencsér (1992, 1994)]). However, in the spirit of modern nonparametric inference, we try to avoid such restrictions on the process structure. Thus, we only assume that we observe a string realization y_1^{n-1} of a zero mean, stationary and ergodic, Gaussian process $\{Y_n\}_{-\infty}^{\infty}$, and try to predict y_n, the value of the process at time n. Note that there are no side information vectors x_1^n in this purely time-series prediction framework.

It is well known for Gaussian time series that the best predictor is a linear function of the past:

$$\mathbb{E}\{Y_n \mid Y_{n-1}, Y_{n-2}, \dots\} = \sum_{j=1}^{\infty} c_j^* Y_{n-j},$$

where the c_j^* minimize the criterion

$$\mathbb{E}\left\{\left(\sum_{j=1}^{\infty} c_j Y_{n-j} - Y_n\right)^2\right\}.$$

Following [Györfi and Lugosi (2001)], we extend the principle of generalized linear estimates to the prediction of Gaussian time series by considering the special case

$$\phi_j^{(k)}(y_{n-k}^{n-1}) = y_{n-j} I_{\{1 \le j \le k\}},$$

i.e.,

$$\tilde{h}_n^{(k)}(y_1^{n-1}) = \sum_{j=1}^{k} c_{n,j} y_{n-j}.$$

Once again, the coefficients $c_{n,j}$ are calculated according to the past observations y_1^{n-1} by minimizing the criterion:

$$\sum_{t=k+1}^{n-1} \left(\sum_{j=1}^{k} c_j y_{t-j} - y_t\right)^2$$

if $n > k$, and the all-zero vector otherwise.

To combine the elementary experts $\tilde{h}^{(k)}$ [Györfi and Lugosi (2001)] applied the so called "doubling-trick", which means that the time axis is segmented into exponentially increasing epochs and at the beginning of each epoch the forecasters' weight is reset.

In this section we propose a much simpler procedure which avoids, in particular, the doubling-trick. To begin, we set

$$h_n^{(k)}(y_1^{n-1}) = T_{\min\{n^\delta, k\}}\left(\tilde{h}_n^{(k)}(y_1^{n-1})\right),$$

where $0 < \delta < \frac{1}{8}$, and combine these experts as before. Precisely, let $\{q_k\}$ be an arbitrarily probability distribution over the positive integers such that for all k, $q_k > 0$, and for $\eta_n > 0$, define the weights

$$w_{k,n} = q_k e^{-\eta_n(n-1)L_{n-1}(h_n^{(k)})}$$

and their normalized values

$$p_{k,n} = \frac{w_{k,n}}{\sum_{i=1}^\infty w_{i,n}}.$$

The prediction strategy g at time n is defined by

$$g_n(y_1^{n-1}) = \sum_{k=1}^\infty p_{k,n} h_n^{(k)}(y_1^{n-1}), \qquad n = 1, 2, \ldots$$

Theorem 5.9. [Biau *et al.* (2010)] *Choose $\eta_t = 1/\sqrt{t}$. Then, the prediction strategy g defined above is universally-consistent with respect to the class of all jointly stationary and ergodic zero-mean Gaussian processes $\{Y_n\}_{-\infty}^\infty$.*

The following corollary shows that the strategy g provides asymptotically a good estimate of the regression function in the following sense:

Corollary 5.3. [Biau *et al.* (2010)] *Under the conditions of Theorem 5.9,*

$$\lim_{n\to\infty} \frac{1}{n} \sum_{t=1}^n \left(\mathbb{E}\{Y_t \mid Y_1^{t-1}\} - g(Y_1^{t-1})\right)^2 = 0 \quad \text{almost surely.}$$

Corollary 5.3 is expressed in terms of an almost sure Cesáro consistency. It is an open problem to know whether there exists a prediction rule g such that

$$\lim_{n\to\infty} \left(\mathbb{E}\{Y_n|Y_1^{n-1}\} - g(Y_1^{n-1})\right) = 0 \quad \text{almost surely} \qquad (5.44)$$

for all stationary and ergodic Gaussian processes. [Schäfer (2002)] proved that, under some additional mild conditions on the Gaussian time series, the consistency (5.44) holds.

References

Algoet, P. (1992). Universal schemes for prediction, gambling, and portfolio selection, *Annals of Probability* **20**, pp. 901–941.

Algoet, P. (1994). The strong law of large numbers for sequential decisions under uncertainity, *IEEE Transactions on Information Theory* **40**, pp. 609–634.

Antos, A., Györfi, L. and György, A. (2005). Individual convergence rates in empirical vector quantizer design, *IEEE Transactions on Information Theory* **51**, pp. 4013–4022.

Auer, P., Cesa-Bianchi, N. and Gentile, C. (2002). Adaptive and self-confident on-line learning algorithms, *Journal of Computer and System Sciences* **64**, 1, pp. 48–75.

Biau, G., Bleakely, K., Györfi, L. and Ottucsák, G. (2010). Nonparametric sequential prediction of time series, *Journal of Nonparametric Statistics* **22**, pp. 297–317.

Breiman, L. (1957). The individual ergodic theorem of information theory, *Annals of Mathematical Statistics* **28**, pp. 809–811, Correction. Breiman L. (1960), Correction notes: Correction to "The individual ergodic theorem of information theory", *Annals of Mathematical Statistics* **31**, pp. 809–810, 1960.

Cesa-Bianchi, N., Freund, Y., Helmbold, D. P., Haussler, D., Schapire, R. and Warmuth, M. K. (1997). How to use expert advice, *Journal of the ACM* **44**, 3, pp. 427–485.

Cesa-Bianchi, N. and Lugosi, G. (2006). *Prediction, Learning, and Games* (Cambridge University Press, Cambridge).

Chow, Y. S. (1965). Local convergence of martingales and the law of large numbers, *Annals of Mathematical Statistics* **36**, pp. 552–558.

Feder, M., Merhav, N. and Gutman, M. (1992). Universal prediction of individual sequences, *IEEE Transactions on Information Theory* **38**, pp. 1258–1270.

Gerencsér, L. (1992). AR(∞) estimation and nonparametric stochastic complexity, *IEEE Transactions on Information Theory* **38**, pp. 1768–1779.

Gerencsér, L. (1994). On Rissanen's predictive stochastic complexity for stationary ARMA processes, *Journal of Statistical Planning and Inference* **41**, pp. 303–325.

Gerencsér, L. and Rissanen, J. (1986). A prediction bound for Gaussian ARMA processes, in *25th IEEE Conference on Decision and Control* (Athens, Greece), pp. 1487–1490.

Györfi, L. (1984). Adaptive linear procedures under general conditions, *IEEE Transactions on Information Theory* **30**, pp. 262–267.

Györfi, L., Kohler, M., Krzyżak, A. and Walk, H. (2002). *A Distribution-Free Theory of Nonparametric Regression* (Springer, New York).

Györfi, L. and Lugosi, G. (2001). Strategies for sequential prediction of stationary time series, in M. Dror, P. L'Ecuyer and F. Szidarovszky (eds.), *Modelling Uncertainty: An Examination of its Theory, Methods and Applications* (Kluwer Academic Publishers, Berlin), pp. 225–248.

Györfi, L., Lugosi, G. and Morvai, G. (1999). A simple randomized algorithm for consistent sequential prediction of ergodic time series, *IEEE Transactions on Information Theory* **45**, pp. 2642–2650.

Györfi, L. and Ottucsák, G. (2007). Sequential prediction of unbounded stationary time series, *IEEE Transactions on Information Theory* **53**, pp. 1866–1872.

Györfi, L. and Wegkamp, M. (2008). Quantization for nonparametric regression, *IEEE Transactions on Information Theory* **54**, pp. 867–874.

Kivinen, J. and Warmuth, M. K. (1999). Averaging expert predictions, in P. F. H. U. Simon (ed.), *Computational Learning Theory: Proceedings of the Fourth European Conference, EuroCOLT'99*, no. 1572 in Lecture Notes in Artificial Intelligence (Springer-Verlag, Berlin), pp. 153–167.

Littlestone, N. and Warmuth, M. K. (1994). The weighted majority algorithm, *Information and Computation* **108**, pp. 212–261.

Merhav, N. and Feder, M. (1998). Universal prediction, *IEEE Transactions on Information Theory* **IT-44**, pp. 2124–2147.

Morvai, G., Yakowitz, S. and Györfi, L. (1996). Nonparametric inference for ergodic, stationary time series, *Annals of Statistics* **24**, pp. 370–379.

Nobel, A. (2003). On optimal sequential prediction for general processes, *IEEE Transactions on Information Theory* **49**, pp. 83–98.

Schäfer, D. (2002). Strongly consistent online forecasting of centered Gaussian processes, *IEEE Transactions on Information Theory* **48**, pp. 791–799.

Singer, A. C. and Feder, M. (1999). Universal linear prediction by model order weighting, *IEEE Transactions on Signal Processing* **47**, pp. 2685–2699.

Singer, A. C. and Feder, M. (2000). Universal linear least-squares prediction, in *Proceedings of the IEEE International Symposium on Information Theory* (Sorrento, Italy), pp. 2354–2362.

Stout, W. F. (1974). *Almost sure convergence* (Academic Press, New York).

Tsay, R. S. (2002). *Analysis of Financial Time Series* (Wiley, New York).

Tsypkin, Y. Z. (1971). *Adaptation and Learning in Automatic Systems* (Academic Press, New York).

Vovk, V. (1990). Aggregating strategies, in *Proceedings of the Third Annual Workshop on Computational Learning Theory* (Morgan Kaufmann, Rochester, NY), pp. 372–383.

Wegkamp, M. (1999). Entropy methods in statistical estimation, *CWI-tract* volume 25, Centrum voor Wiskunde en Informatica, Amterdam.

Yang, Y. (2000). Combining different procedures for adaptive regression, *Journal of Multivariate Analysis* **74**, pp. 135–161.

Chapter 6

Empirical Pricing American Put Options

László Györfi* and András Telcs[†]

Department of Computer Science and Information Theory,
Budapest University of Technology and Economics,
H-1117, Magyar tudósok körútja 2., Budapest, Hungary.
*gyorfi@cs.bme.hu, [†]telcs@cs.bme.hu

In this chapter, we study the empirical-pricing of American options. The pricing of American options is an optimal stopping problem, which can be derived from a backward recursion such that, in each step of the recursion, one needs conditional expectations. For empirical-pricing, [Longstaff and Schwartz (2001)] suggested replacing the conditional expectations by regression function estimates. We survey the current literature and the main techniques of nonparametric regression estimates, and derive new empirical-pricing algorithms.

6.1. Introduction: The Valuation of Option Price

One of the most important problems in option-pricing theory is the valuation and optimal exercise of derivatives with American-style exercise features. Such derivatives are, for example, the equity, commodity, foreign exchange, insurance, energy, municipal, mortgage, credit, convertible, swap, emerging markets, etc. Despite recent progress, the valuation and optimal exercise of American options remains one of the most challenging problems in derivatives finance. Many financial instruments allow the contract to be exercised early, before expiry. For example, many exchange traded options are of the American type and allow the holder any exercise date before expiry, mortgages often have embedded prepayment options such that the mortgage can be amortized or repayed, and life-insurance contracts often allow for early surrender. In this chapter, we consider data-driven pricing of options with early-exercise features.

6.1.1. Notations

Let X_t be the asset price at time t, K the strike price, and r the discount rate. For an American put option, the payoff function f_t with discount factor e^{-rt} is

$$f_t(X_t) = e^{-rt} (K - X_t)^+ \ .$$

For maturity time T, let $\mathbb{T} = \{1, \dots, T\}$ be the time frame for American options. Let $\mathcal{F}_t$ denote the σ-algebra generated by $X_0 = 1, X_1, \dots, X_t$; then, an integer-valued random variable τ is called the stopping time if $\{\tau = t\} \in \mathcal{F}_t$ for all $t = 1, \dots, T$. If $\widetilde{\mathfrak{T}}(0, \dots, T)$ stands for the set of stopping times taking values in $(0, \dots, T)$, then the task of pricing the American option is to determine

$$V_0 = \sup_{\tau \in \widetilde{\mathfrak{T}}(0,\dots,T)} \mathbb{E}\{f_\tau(X_\tau)\}. \qquad (6.1)$$

The main principles of pricing American put options described below can be extended to more general payoffs. For example, the payoffs may depend on many assets' prices (cf. [Tsitsiklis and Roy (2001)]).

Let τ^* be the optimum stopping time, i.e.,

$$\mathbb{E}\{f_{\tau^*}(X_{\tau^*})\} = \sup_{\tau \in \widetilde{\mathfrak{T}}(0,\dots,T)} \mathbb{E}\{f_\tau(X_\tau)\}$$

6.1.2. Optimal Stopping

An alternative formulation of τ^* can be derived as follows. Introduce $q_t(x)$ continuation value, that is,

$$q_t(x) = \sup_{\tau \in \widetilde{\mathfrak{T}}\{t+1,\dots,T\}} \mathbb{E}\{f_\tau(X_\tau) \mid X_t = x\}, \qquad (6.2)$$

where $\widetilde{\mathfrak{T}}\{t+1, \dots, T\}$ refers to the possible stopping times taking values in $\{t+1, \dots, T\}$.

Theorem 6.1. (cf. [Chow et al. (1971)], [Shiryaev (1978)] and [Kohler (2010)]) *Put*

$$\tau^q = \min\{1 \leq s \leq T : q_s(X_s) \leq f_s(X_s)\}.$$

If the asset's prices $\{X_t\}$ form a Markov process, then

$$\tau^* = \tau^q.$$

The reasoning behind the optimal stopping rule τ^q is that, at any exercise time, the holder of an American option optimally compares the payoff from immediate exercise with the expected payoff from continuation, and then exercises if the immediate payoff is higher. Thus, the optimal exercise strategy is fundamentally determined by the conditional expectation of the payoff from continuing to keep the option alive. The key insight underlying the current approaches is that this conditional expectation can be estimated from data.

As a byproduct of the proof of Theorem 6.1, one may check the following:

Theorem 6.2. (cf. [Tsitsiklis and Roy (1999)] and [Kohler (2010)]) *We find that*

$$q_T(x) = 0,$$

while, at any $t < T$,

$$q_t(x) = \mathbb{E}\left\{\max\left\{f_{t+1}\left(X_{t+1}\right), q_{t+1}\left(X_{t+1}\right)\right\} \mid X_t = x\right\}, \qquad (6.3)$$

which means that there is a backward-recursive scheme.

(6.3) implies that

$$
\begin{aligned}
q_t(x) &\\
&= \mathbb{E}\left\{\max\left\{f_{t+1}\left(X_{t+1}\right), q_{t+1}\left(X_{t+1}\right)\right\} \mid X_t = x\right\} \\
&= \mathbb{E}\left\{\max\left\{e^{-r(t+1)}\left(K - X_{t+1}\right)^+, q_{t+1}\left(X_{t+1}\right)\right\} \mid X_t = x\right\} \\
&= \mathbb{E}\left\{\max\left\{e^{-r(t+1)}\left(K - \frac{X_{t+1}}{X_t}X_t\right)^+, q_{t+1}\left(\frac{X_{t+1}}{X_t}X_t\right)\right\} \Big| X_t = x\right\} \\
&= \mathbb{E}\left\{\max\left\{e^{-r(t+1)}\left(K - \frac{X_{t+1}}{X_t}x\right)^+, q_{t+1}\left(\frac{X_{t+1}}{X_t}x\right)\right\} \Big| X_t = x\right\}.
\end{aligned}
$$

$$(6.4)$$

6.1.3. *Martingale Approach: The Primal-Dual Problem*

As we defined in the introduction, the initial problem is to find the optimal stopping time which provides the price of an American option:

$$V_0 = \sup_{\tau \in \widetilde{\mathfrak{T}}(0,\ldots,T)} \mathbb{E}\left\{f_\tau\left(X_\tau\right)\right\},$$

where the sup is taken over the stopping times τ. The dual problem is formulated by [Rogers (2002)] and [Haugh and Kogan (2004)] to obtain an alternative valuation method. Let

$$U_0 = \inf_{M \in \mathcal{M}} \mathbb{E} \left\{ \max_{t \in \{0,1,\dots T\}} (f_t(X_t) - M_t) \right\}, \qquad (6.5)$$

where $\mathcal{M}$ is the set of martingales with $M_0 = 0$ and with the same filtration $\sigma(X_t, \dots, X_1)$. The dual method is based on the next theorem.

Theorem 6.3. (cf. [Rogers (2002)], [Haugh and Kogan (2004)], [Glasserman (2004)] and [Kohler (2010)])
If X_t is a Markov process, then

$$U_0 = V_0$$

This result is based on the important observation that one can obtain a martingale from the pay-off function and continuation value in a natural way.

Theorem 6.4. (cf. [Glasserman (2004)], [Tsitsiklis and Roy (1999)] and [Kohler (2010)])
The optimal martingale is of form

$$M_t^* = \sum_{s=1}^{t} \left(\max \left\{ f_s(X_s), q_s(X_s) \right\} - q_{s-1}(X_{s-1}) \right)$$

and, indeed, M_t^* is a martingale.

The valuation task now is converted into an estimate of the martingale M_t^*.

6.1.4. *Lower and Upper Bounds of $q_t(x)$*

In pricing American options, the continuation values $q_t(x)$ play an important role. For empirical-pricing, one has to estimate them, which is possible using the backward recursion (6.3). However, using this recursion, the estimation errors are accumulated, therefore, there is a need to control the error propagation.

We introduce a lower bound of $q_t(x)$

$$q_t^{(l)}(x) = \max_{s \in \{t+1, \dots, T\}} \mathbb{E} \left\{ f_s(X_s) | X_t = x \right\}.$$

Since any constant $\tau = s$ is a stopping time, we have that

$$q_t^{(l)}(x) \le q_t(x).$$

We shall show that $q_t^{(l)}(x)$ can be estimated more easily than that of $q_t(x)$ and the estimate has a fast rate of convergence, so, if $q_{t,n}^{(l)}(x)$ and $q_{t,n}(x)$ are the estimates of $q_t^{(l)}(x)$ and $q_t(x)$, resp., then

$$\hat{q}_{t,n}(x) \triangleq \max\{q_{t,n}(x), q_{t,n}^{(l)}(x)\}$$

is a hopefully improved estimate of $q_t(x)$.

Next, we introduce an upper bound. For $\tau \in \tilde{\mathfrak{T}}\{t+1, \ldots, T\}$, we have that

$$f_\tau(X_\tau) \leq \max_{s \in \{t+1, \ldots, T\}} f_s(X_s).$$

Therefore,

$$q_t(x) = \sup_{\tau \in \tilde{\mathfrak{T}}\{t+1, \ldots, T\}} \mathbb{E}\left\{f_\tau(X_\tau) \mid X_t = x\right\}$$

$$\leq \mathbb{E}\left\{\max_{s \in \{t+1, \ldots, T\}} f_s(X_s) \mid X_t = x\right\}.$$

Introduce the notation

$$q_t^{(u)}(x) \triangleq \mathbb{E}\left\{\max_{s \in \{t+1, \ldots, T\}} f_s(X_s) \mid X_t = x\right\},$$

then we obtain an upper bound

$$q_t(x) \leq q_t^{(u)}(x).$$

Again, $q_t^{(u)}(x)$ can be estimated more easily than that of $q_t(x)$ and the estimate has a fast rate of convergence, so, if $q_{t,n}^{(u)}(x)$ and $q_{t,n}(x)$ are the estimates of $q_t^{(u)}(x)$ and $q_t(x)$, resp., then

$$\hat{q}_{t,n}(x) \triangleq \min\{q_{t,n}(x), q_{t,n}^{(u)}(x)\}$$

is an improved estimate of $q_t(x)$.

The combination of the lower and upper bounds reads as follows:

$$\max_{s \in \{t+1, \ldots, T\}} \mathbb{E}\left\{f_s(X_s)|X_t = x\right\} \leq q_t(x) \leq \mathbb{E}\left\{\max_{s \in \{t+1, \ldots, T\}} f_s(X_s) \mid X_t = x\right\},$$

while the improved estimate has the form

$$\hat{q}_{t,n}(x) = \begin{cases} q_{t,n}^{(u)}(x) \text{ if } & q_{t,n}^{(u)}(x) < q_{t,n}(x), \\ q_{t,n}(x) \text{ if } & q_{t,n}^{(u)}(x) \geq q_{t,n}(x) \geq q_{t,n}^{(l)}(x), \\ q_{t,n}^{(l)}(x) \text{ if } & q_{t,n}(x) < q_{t,n}^{(l)}(x). \end{cases}$$

6.1.5. *Sampling*

In a real-life problem, we have a single, historical data sequence $X_1, \ldots, X_N$.

Definition 6.1. The process $\{X_t\}$ is called memoryless multiplicative increments, if $X_1/X_0, X_2/X_1, \ldots$ are independent random variables.

Definition 6.2. The process $\{X_t\}$ is called of stationary multiplicative increments, if the sequence $X_1/X_0, X_2/X_1, \ldots$ is strictly stationary.

As mentioned earlier, the continuation value $q_t(x)$ plays an important role in optimum pricing, which is the supremum of conditional expectations. Conditional expectations can be considered as regression functions, and, in empirical-pricing, the regression function is replaced by its estimate. For regression function estimation, we are given independent and identically distributed (i.i.d) copies of $X_1, \ldots, X_T$, i.e., one generates i.i.d. sample path prices

$$X_{i,1}, \ldots, X_{i,T}, \tag{6.6}$$

$i = 1, \ldots, n$.

Based on the historical data sequence $X_1, \ldots, X_N$, one can construct samples for (6.6) as follows:

(i) For Monte Carlo sampling, one assumes that the data-generating process is completely known, i.e., that there is a perfect parametric model and all parameters of this process are already estimated from historical data $X_1, \ldots, X_N$ (cf. [Longstaff and Schwartz (2001)]). Thus, one can artificially generate independent sample paths (6.6). The weakness of this approach is that, usually, the size N of the historical data is not large enough in order to have a good model and reliable parameter estimates.

(ii) For disjoint sampling, $N = nT$ and $i = 1, \ldots, n = N/T$. However, we do not have the required i.i.d. property unless the process $X_1, \ldots, X_{nT}$ has memoryless and stationary multiplicative increments, which means that $X_1/X_0, X_2/X_1, \ldots, X_{nT}/X_{nT-1}$ are i.i.d.

(iii) For sliding sampling,

$$X_{i,t} \triangleq \frac{X_{i+t}}{X_i}, \tag{6.7}$$

$i = 1, \ldots, n = N - T$. In this way we obtain a large sample; however, there is no i.i.d. property.

(iv) For bootstrap sampling, we generate i.i.d. random variables $T_1, \ldots, T_n$ uniformly distributed on $1, \ldots, N - T$ and

$$X_{i,t} \triangleq \frac{X_{T_i+t}}{X_{T_i}}, \tag{6.8}$$

$i = 1, \ldots, n.$

6.1.6. *Empirical-Pricing and Optimal Exercising of American Options*

If the continuation values $q_t(x)$, $t = 1, \ldots T$ were known, then the *optimal stopping time* τ_i for path $X_{i,1}, \ldots, X_{i,T}$ can be calculated:

$$\tau_i = \min \{1 \le s \le T : q_s(X_{i,s}) \le f_s(X_{i,s})\}.$$

Then, the price V_0 can be estimated by the average

$$\frac{1}{n} \sum_{i=1}^{n} f_{\tau_i}(X_{\tau_i}). \tag{6.9}$$

The continuation values $q_t(x)$, $t = 1, \ldots, T$ are unknown; therefore, one has to generate some estimates $q_{t,n}(x)$, $t = 1, \ldots, T$. [Kohler *et al.* (2008)] suggested a splitting approach as follows. Split the sample $\{X_{i,1}, \ldots, X_{i,T}\}_{i=1}^{n}$ into two samples: $\{X_{i,1}, \ldots, X_{i,T}\}_{i=1}^{m}$ and $\{X_{i,1}, \ldots, X_{i,T}\}_{i=m+1}^{n}$. We estimate $q_t(x)$ by $q_{t,m}(x)$, $(t = 1, \ldots, T)$ from $\{X_{i,1}, \ldots, X_{i,T}\}_{i=1}^{m}$ and construct some approximations of the optimal stopping time τ_i for path $X_{i,1}, \ldots, X_{i,T}$

$$\tau_{i,m} = \min \{1 \le s \le T : q_{s,m}(X_{i,s}) \le f_s(X_{i,s})\},$$

and, then, the price V_0 can be estimated by the average

$$\frac{1}{n-m} \sum_{i=m+1}^{n} f_{\tau_{i,m}}(X_{\tau_{i,m}}).$$

For empirical exercising at the time frame $[N+1, N+T]$, we are given the past data $X_1, \ldots, X_N$ based on which we generate some estimates $q_{t,N}(x)$, $t = 1, \ldots, T$. Then, the *empirical exercising of an American option* can be defined by the stopping time

$$\tau_N = \min \{1 \le s \le T : q_{s,N}(X_{N+s}/X_N) \le f_s(X_{N+s}/X_N)\}.$$

If the continuation values $q_t(x)$, $t = 1, \ldots, T$ were known, then the *optimal martingale* $M_{i,t}^*$ for path $X_{i,1}, \ldots, X_{i,T}$ can be calculated:

$$M_{i,t}^* = \sum_{s=1}^{t} (\max \{f_s(X_{i,s}), q_s(X_{i,s})\} - q_{s-1}(X_{i,s-1})).$$

Then, the price V_0 can be estimated by the average

$$\frac{1}{n} \sum_{i=1}^{n} \max_{t \in \{0,1....T\}} \left(f_t \left(X_{i,t} \right) - M_{i,t}^* \right). \tag{6.10}$$

The continuation values $q_t(x)$, $t = 1, \ldots, T$ are unknown; then, using the splitting approach described above, generate some estimates $q_{t,m}(x)$, $t = 1, \ldots, T$ and the approximations of the optimal martingale $M_{i,t}^*$ for path $X_{i,1}, \ldots, X_{i,T}$:

$$M_{i,t,m}^* = \sum_{s=1}^{t} \left(\max \left\{ f_s \left(X_{i,s} \right), q_{s,m} \left(X_{i,s} \right) \right\} - q_{s-1,m} \left(X_{i,s-1} \right) \right).$$

Then, the price V_0 can be estimated by the average

$$V_{0,n} = \frac{1}{n - m} \sum_{i=m+1}^{n} \max_{t \in \{0,1,\ldots,T\}} \left(f_t \left(X_{i,t} \right) - M_{i,t,m}^* \right).$$

For option-pricing, a nonparametric estimation scheme was first proposed by [Carrier (1996)], while [Tsitsiklis and Roy (1999)] and [Longstaff and Schwartz (2001)] estimated the continuation value.

6.2. Special Case: Pricing for a Process with Memoryless and Stationary Multiplicative Increments

In this section, we assume that the asset's prices $\{X_t\}$ have memoryless and stationary multiplicative increments. This properties imply that, for $s > t$, $\frac{X_s}{X_t}$ and X_t are independent, and $\frac{X_s}{X_t}$ and $\frac{X_{s-t}}{X_0} = X_{s-t}$ have the same distribution.

6.2.1. *Estimating q_t*

For $t < T$, the recursion (6.4) implies that

$$
\begin{aligned}
q_t(x) &= \mathbb{E} \left\{ \max \left\{ f_{t+1} \left(X_{t+1} \right), q_{t+1} \left(X_{t+1} \right) \right\} \mid X_t = x \right\} \\
&= \mathbb{E} \left\{ \max \left\{ e^{-r(t+1)} \left(K - \frac{X_{t+1}}{X_t} x \right)^+, q_{t+1} \left(\frac{X_{t+1}}{X_t} x \right) \right\} \Big| X_t = x \right\} \\
&= \mathbb{E} \left\{ \max \left\{ e^{-r(t+1)} \left(K - \frac{X_{t+1}}{X_t} x \right)^+, q_{t+1} \left(\frac{X_{t+1}}{X_t} x \right) \right\} \right\} \\
&= \mathbb{E} \left\{ \max \left\{ e^{-r(t+1)} \left(K - X_1 x \right)^+, q_{t+1} \left(X_1 x \right) \right\} \right\}, \tag{6.11}
\end{aligned}
$$

where, in the last two steps, we assumed independent and stationary multiplicative increments. By a backward induction, we find that, for fixed t, $q_t(x)$ is a monotonically-decreasing and convex function of x.

If we are given data $X_1, \ldots, X_N$, $i = 1, \ldots, N$ then, for any fixed t, let $q_{t+1,N}(x)$ be an estimate of $q_{t+1}(x)$. Thus, introduce the estimate of $q_t(x)$ in a backward recursive way as follows:

$$q_{t,N}(x) = \frac{1}{N} \sum_{i=1}^{N} \max \left\{ e^{-r(t+1)} \left(K - xX_i/X_{i-1} \right)^+, q_{t+1,N}(xX_i/X_{i-1}) \right\}.$$

$$(6.12)$$

From (6.12) we can derive a numerical procedure by considering a grid

$$\mathbf{G} \triangleq \{j \cdot h\},$$

$j = 1, 2, \ldots$, where the step size of the grid $h > 0$, for example $h = 0.01$. In each step of (6.12) we make the recursion for $x \in \mathbf{G}$, and then linearly interpolate for $x \notin \mathbf{G}$.

A possible weakness of this estimate is that estimation errors are cumulated. Therefore, we also consider the estimates of the lower and upper bounds.

6.2.2. *Estimating the Lower and Upper Bounds of $q_t(x)$*

For a memoryless process, the lower bound of $q_t(x)$ has a simple form

$$q_t^{(l)}(x) = \max_{s \in \{t+1,\ldots,T\}} \mathbb{E}\left\{ f_s(X_s) | X_t = x \right\}$$

$$= \max_{s \in \{t+1,\ldots,T\}} e^{-rs} \mathbb{E}\left\{ \left(K - \frac{X_s}{X_t} X_t \right)^+ \Big| X_t = x \right\}$$

$$= \max_{s \in \{t+1,\ldots,T\}} e^{-rs} \mathbb{E}\left\{ \left(K - \frac{X_s}{X_t} x \right)^+ \Big| X_t = x \right\}$$

$$= \max_{s \in \{t+1,\ldots,T\}} e^{-rs} \mathbb{E}\left\{ \left(K - \frac{X_s}{X_t} x \right)^+ \right\}$$

$$= \max_{s \in \{t+1,\ldots,T\}} e^{-rs} \mathbb{E}\left\{ (K - X_{s-t} x)^+ \right\},$$

where, in the last two steps, we assumed memoryless and stationary multiplicative increments.

Thus,

$$q_t^{(l)}(x) = \sup_{s \in \{t+1,\ldots,T\}} e^{-rs} \mathbb{E}\left\{ (K - X_{s-t} x)^+ \right\}.$$

If we are given data $X_{i,1}, \ldots, X_{i,T}$, $i = 1, \ldots n$, then the estimate of $q_t^{(l)}(x)$ would be

$$q_{t,n}^{(l)}(x) = \max_{s \in \{t+1,\ldots,T\}} e^{-rs} \frac{1}{n} \sum_{i=1}^{n} (K - X_{i,s-t}x)^+.$$

Concerning the upper bound, the previous arguments imply that

$$q_t^{(u)}(x) = \mathbb{E}\left\{ \max_{s \in \{t+1,\ldots,T\}} f_s(X_s) \mid X_t = x \right\}$$

$$= \mathbb{E}\left\{ \max_{s \in \{t+1,\ldots,T\}} e^{-rs}(K - X_s)^+ \Big| X_t = x \right\}$$

$$= \mathbb{E}\left\{ \max_{s \in \{t+1,\ldots,T\}} e^{-rs}\left(K - \frac{X_s}{X_t}X_t\right)^+ \Big| X_t = x \right\}$$

$$= \mathbb{E}\left\{ \max_{s \in \{t+1,\ldots,T\}} e^{-rs}\left(K - \frac{X_s}{X_t}x\right)^+ \Big| X_t = x \right\}$$

$$= \mathbb{E}\left\{ \max_{s \in \{t+1,\ldots,T\}} e^{-rs}\left(K - \frac{X_s}{X_t}x\right)^+ \right\}$$

$$= \mathbb{E}\left\{ \max_{s \in \{t+1,\ldots,T\}} e^{-rs}(K - X_{s-t}x)^+ \right\}.$$

If we are given data $X_{i,1}, \ldots, X_{i,T}$, $i = 1, \ldots n$, then the estimate of $q_t^{(u)}(x)$ would be

$$q_{t,n}^{(u)}(x) = \frac{1}{n} \sum_{i=1}^{n} \max_{s \in \{t+1,\ldots,T\}} e^{-rs}(K - X_{i,s-t}x)^+.$$

The combination of the lower an upper bounds reads as follows:

$$\max_{s \in \{t+1,\ldots,T\}} \mathbb{E}\left\{ e^{-rs}(K - X_{s-t}x)^+ \right\}$$

$$\le q_t(x) \le \mathbb{E}\left\{ \max_{s \in \{t+1,\ldots,T\}} e^{-rs}(K - X_{s-t}x)^+ \right\}.$$

Again, using the estimates of the lower and upper bounds, we suggest a truncation of the estimates of the continuation value:

$$\hat{q}_{t,N}(x) = \begin{cases} q_{t,n}^{(u)}(x) & \text{if } q_{t,n}^{(u)}(x) < q_{t,N}(x), \\ q_{t,N}(x) & \text{if } q_{t,n}^{(u)}(x) \ge q_{t,N}(x) \ge q_{t,n}^{(l)}(x), \\ q_{t,n}^{(l)}(x) & \text{if } q_{t,N}(x) < q_{t,n}^{(l)}(x). \end{cases}$$

6.2.3. *The Growth Rate of an Asset and the Black–Scholes Model*

In this section, we still assume that the asset's prices $\{X_t\}$ have memoryless and stationary multiplicative increments, and, in discrete time, show that the Black–Scholes formula results in a good approximation of the lower bound $q_t^{(l)}(x)$. Consider an asset, the evolution of which is characterized by its price X_t at trading period (let us say trading day) t. In order to normalize, put $X_0 = 1$. X_t has an exponential trend

$$X_t = e^{tW_t} \approx e^{tW},$$

with average growth rate (average daily yield)

$$W_t \triangleq \frac{1}{t} \ln X_t$$

and with asymptotic average growth rate

$$W \triangleq \lim_{t \to \infty} \frac{1}{t} \ln X_t.$$

Introduce the returns Z_t as follows:

$$Z_t = \frac{X_t}{X_{t-1}}.$$

Thus, the return Z_t denotes the amount obtained after investing unit capital in the asset on the t-th trading period. Because $\{X_t\}$ is of independent and stationary multiplicative increments, the sequence $\{Z_t\}$ is i.i.d. Then, the strong law of large numbers (cf. [Stout (1974)]) implies that

$$
\begin{aligned}
W_t &= \frac{1}{t} \ln X_t \\
&= \frac{1}{t} \ln \prod_{i=1}^{t} \frac{X_i}{X_{i-1}} \\
&= \frac{1}{n} \ln \prod_{i=1}^{n} Z_i \\
&= \frac{1}{n} \sum_{i=1}^{n} \ln Z_i \\
&\to \mathbb{E}\{\ln Z_1\} = \mathbb{E}\{\ln X_1\}
\end{aligned}
$$

almost surely, therefore,

$$W = \mathbb{E}\{\ln X_1\}.$$

The problem is how to calculate $\mathbb{E}\{\ln X_1\}$. It is not an easy task, as one must know the distribution of X_1. For the approximate calculation of the log-optimal portfolio, [Vajda (2006)] suggested using the second-order Taylor expansion of the function $\ln z$ at $z = 1$:

$$h(z) \triangleq z - 1 - \frac{1}{2}(z - 1)^2.$$

For daily returns, this is a very good approximation, so it is a natural idea to introduce the semi-log approximation of the asymptotic growth rate:

$$\widetilde{W} = \mathbb{E}\{h(X_1)\}.$$

$\widetilde{W}$ has the advantage that it can be calculated just knowing the first and second moments of X_1. Put

$$\mathbb{E}\{X_1\} = 1 + r_a$$

and

$$\mathbf{Var}(X_1) = \sigma^2,$$

then,

$$\widetilde{W} = \mathbb{E}\{h(X_1)\} = \mathbb{E}\{X_1 - 1 - \frac{1}{2}(X_1 - 1)^2\} = r_a - \frac{\sigma^2 + r_a^2}{2} \approx r_a - \frac{\sigma^2}{2}.$$

Table 6.1 summarizes the growth rate of some large stocks on the New York Stock Exchange (NYSE). The database used contains daily relative closing prices of several stocks and it is normalized by dividend and splits for all trading days. For more information about the database, see the webpage [Gelencsér and Ottucsák (2006)]. One can see that $\widetilde{W}$ is a very good approximation of W.

If the expiration time T is much larger than one day then, for $\ln X_T$, we cannot apply the semi-log approximation, and we should instead approximate the distribution of $\ln X_T$.

As for the binomial model, Cox–Ross–Rubinstein model or for the construction of geometric Brownian motion (cf. [Luenberger (1998)]), we assumed that $\{Z_t\}$ are i.i.d. Then,

$$
\mathbf{Var}\left(\sum_{i=1}^{t} \ln Z_i\right) \approx \mathbf{Var}\left(\sum_{i=1}^{t} h(Z_i)\right) = t\mathbf{Var}\left(h(Z_1)\right)
$$

$$
= t\mathbf{Var}\left(X_1 - 1 - \frac{1}{2}(X_1 - 1)^2\right)
$$

$$
= t\Big(\mathbb{E}\{(X_1 - 1)^2\} - \mathbb{E}\{(X_1 - 1)^3\}
$$

$$
+ \frac{1}{4}\mathbb{E}\{(X_1 - 1)^4\} - (r_a - \frac{1}{2}(\sigma^2 + r_a^2))^2\Big)
$$

$$
\approx t\sigma^2.
$$

Thus, by the central limit theorem, we find that $\ln X_t$ is approximately Gaussian distributed with mean $t(r_a - (\sigma^2 + r_a^2)/2) \approx t(r_a - \sigma^2/2)$ and variance $t\sigma^2$:

$$
\ln X_t \stackrel{\mathcal{D}}{\approx} \mathcal{N}\left(t(r_a - \sigma^2/2), t\sigma^2\right),
$$

so we have derived the discrete-time version of the Black–Scholes model.

Table 6.1.   The average empirical daily yield, variance, growth rate and estimated growth rate for the 19 stocks from [Gelencsér and Ottucsák (2006)].

STOCK	r_a	σ	W	$\widetilde{W}$
AHP	0.000602	0.0160	0.000473	0.000474
ALCOA	0.000516	0.0185	0.000343	0.000343
AMERB	0.000616	0.0145	0.000511	0.000510
COKE	0.000645	0.0152	0.000528	0.000528
DOW	0.000576	0.0167	0.000436	0.000436
DUPONT	0.000442	0.0153	0.000325	0.000324
FORD	0.000526	0.0184	0.000356	0.000356
GE	0.000591	0.0151	0.000476	0.000476
GM	0.000408	0.0171	0.000261	0.000261
HP	0.000807	0.0227	0.000548	0.000548
IBM	0.000495	0.0161	0.000365	0.000365
INGER	0.000571	0.0177	0.000413	0.000413
JNJ	0.000712	0.0153	0.000593	0.000593
KIMBC	0.000599	0.0154	0.000479	0.000480
MERCK	0.000669	0.0156	0.000546	0.000546
MMM	0.000513	0.0144	0.000408	0.000408
MORRIS	0.000874	0.0169	0.000729	0.000730
PANDG	0.000579	0.0140	0.000478	0.000479
SCHLUM	0.000741	0.0191	0.000557	0.000557

We have that

$$\ln X_t \stackrel{\mathcal{D}}{\approx} \mathcal{N}\left(tv_0, t\sigma^2\right),$$

where

$$v_0 = r_a - \sigma^2/2.$$

Let $Z \stackrel{\mathcal{D}}{=} \mathcal{N}(0,1)$, then

$$\mathbb{E}\left\{(K - xX_t)^+\right\} = \mathbb{E}\left\{\left(K - xe^{\ln X_t}\right)^+\right\}$$
$$= \int_{-\infty}^{\infty} \left(K - xe^{tv_0 + \sqrt{t}\sigma z}\right)^+ \frac{1}{\sqrt{2\pi}\sigma} e^{-\frac{z^2}{2\sigma^2}} dz.$$

We have

$$K - xe^{tv_0 + \sqrt{t}\sigma z} > 0$$

if and only if

$$\log \frac{K}{x} > tv_0 + \sqrt{t}\sigma z,$$

equivalently

$$z_0 \triangleq \frac{\log \frac{K}{x} - tv_0}{\sqrt{t}\sigma} > z.$$

Thus,

$$\mathbb{E}\left\{(K - xX_t)^+\right\} = \int_{-\infty}^{z_0} \left(K - xe^{tv_0 + \sqrt{t}\sigma z}\right)^+ \frac{1}{\sqrt{2\pi}\sigma} e^{-\frac{z^2}{2\sigma^2}} dz$$
$$= K\Phi(z_0) - \frac{xe^{tv_0}}{\sqrt{2\pi}} \int_{-\infty}^{z_0} e^{\sqrt{t}\sigma z - z_0^2/2} dz$$
$$= K\Phi(z_0) - \frac{xe^{t(v_0 + \sigma^2/2)}}{\sqrt{2\pi}} \int_{-\infty}^{z_0} e^{\frac{(z - \sqrt{t}\sigma)^2}{2}} dz$$
$$= K\Phi(z_0) - xe^{t(v_0 + \sigma^2/2)}\Phi\left(z_0 - \sqrt{t}\sigma\right).$$

Consequently,

$$e^{-rt}\mathbb{E}\left\{(K - xX_t)^+\right\}$$
$$= e^{-rt}\left(K\Phi\left(\frac{\log \frac{K}{x} - tv_0}{\sqrt{t}\sigma}\right) - xe^{t(v_0 + \sigma^2/2)}\Phi\left(\frac{\log \frac{K}{x} - tv_0}{\sqrt{t}\sigma} - \sqrt{t}\sigma\right)\right),$$

therefore, we find that

$$q_t^{(l)}(x)$$

$$= \sup_{s \in \{t+1,\dots,T\}} e^{-rs} \mathbb{E} \left\{ (K - X_{s-t} x)^+ \right\}$$

$$= e^{-rt} \cdot \sup_{s \in \{1,\dots,T-t\}} e^{-rs} \left(K \Phi \left(\frac{\log \frac{K}{x} - s v_0}{\sqrt{s}\sigma} \right) \right.$$

$$\left. - x e^{s(v_0 + \sigma^2/2)} \Phi \left(\frac{\log \frac{K}{x} - s v_0 - s\sigma^2}{\sqrt{s}\sigma} \right) \right).$$

6.3. Nonparametric Regression Estimation

In order to introduce efficient estimates of $q_t(x)$ for general Markov processes, we briefly summarize the basics of nonparametric regression estimation. In regression analysis, one considers a random vector (X, Y), where X and Y are $\mathbb{R}$-valued, and one is interested in how the value of the so-called response variable Y depends on the value of the observation X. This means that one wants to find a function $f : \mathbb{R} \to \mathbb{R}$, such that $f(X)$ is a "good approximation of Y", that is, $f(X)$ should be close to Y in some sense, which is equivalent to making $|f(X) - Y|$ "small". Since X and Y are random, $|f(X) - Y|$ is random as well, therefore, it is not clear what "small $|f(X) - Y|$" means. We can resolve this problem by introducing the so-called *mean-squared error* of f,

$$\mathbb{E}|f(X) - Y|^2,$$

and requiring it to be as small as possible. Therefore, we are interested in a function $m : \mathbb{R} \to \mathbb{R}$ such that

$$\mathbb{E}|m(X) - Y|^2 = \min_{f:\mathbb{R}\to\mathbb{R}} \mathbb{E}|f(X) - Y|^2.$$

According to Chapter 5 of this volume, such a function can be obtained explicitly by the *regression function*:

$$m(x) = \mathbb{E}\{Y|X = x\}.$$

In applications, the distribution of (X, Y) (and, hence, also the regression function) is usually unknown. Therefore, it is impossible to predict Y using $m(X)$. However, it is often possible to observe data according to the distribution of (X, Y) and to estimate the regression function from these data.

To be more precise, denote by (X, Y), (X_1, Y_1), (X_2, Y_2), ... i.i.d. random variables with $\mathbb{E}Y^2 < \infty$. Let $\mathcal{D}_n$ be the set of *data* defined by

$$\mathcal{D}_n = \{(X_1, Y_1), \ldots, (X_n, Y_n)\} \,.$$

In the regression function estimation problem one wants to use the data $\mathcal{D}_n$ in order to construct an estimate $m_n : \mathbb{R} \to \mathbb{R}$ of the regression function m. Here $m_n(x) = m_n(x, \mathcal{D}_n)$ is a measurable function of x and the data. For simplicity, we will suppress $\mathcal{D}_n$ in the notation and write $m_n(x)$ instead of $m_n(x, \mathcal{D}_n)$.

In this section, we describe the basic principles of nonparametric regression estimation: *local averaging*, or *least-squares estimation*. (Concerning the details, see Chapter 5 of this volume and [Györfi *et al.* (2002)].)

The local-averaging estimates of $m(x)$ can be written as

$$m_n(x) = \sum_{i=1}^{n} W_{n,i}(x) \cdot Y_i,$$

where the weights $W_{n,i}(x) = W_{n,i}(x, X_1, \ldots, X_n) \in \mathbb{R}$ depend on $X_1, \ldots, X_n$. Usually, the weights are nonnegative and $W_{n,i}(x)$ is "small" if X_i is "far" from x.

An example of such an estimate is the *partitioning estimate*. Here, one chooses a finite or countably-infinite partition $\mathcal{P}_n = \{A_{n,1}, A_{n,2}, \ldots\}$ of $\mathbb{R}$ consisting of cells $A_{n,j} \subseteq \mathbb{R}$ and defines, for $x \in A_{n,j}$, the estimate by averaging Y_i's with the corresponding X_i's in $A_{n,j}$, i.e.,

$$m_n(x) = \frac{\sum_{i=1}^{n} I_{\{X_i \in A_{n,j}\}} Y_i}{\sum_{i=1}^{n} I_{\{X_i \in A_{n,j}\}}} \quad \text{for } x \in A_{n,j},$$

where I_A denotes the indicator function of set A. Here, and in the following, we use the convention $\frac{0}{0} = 0$. For the partition $\mathcal{P}_n$, the most important example is when the cells $A_{n,j}$ are intervals of length h_n. For interval partition, the consistency conditions mean that

$$\lim_{n \to \infty} h_n = 0 \quad \text{and} \quad \lim_{n \to \infty} n h_n = \infty. \qquad (6.13)$$

The second example of a local averaging estimate is the *Nadaraya–Watson kernel estimate*. Let $K : \mathbb{R} \to \mathbb{R}_+$ be a function called the kernel function, and let $h > 0$ be a bandwidth. The kernel estimate is defined by

$$m_n(x) = \frac{\sum_{i=1}^{n} K\left(\frac{x - X_i}{h}\right) Y_i}{\sum_{i=1}^{n} K\left(\frac{x - X_i}{h}\right)}.$$

Here the estimate is a weighted average of the Y_i, where the weight of Y_i (i.e., the influence of Y_i on the value of the estimate at x) depends on the distance between X_i and x. For the bandwidth $h = h_n$, the consistency conditions are (6.13). If one uses the so-called naïve kernel (or window kernel) $K(x) = I_{\{\|x\| \leq 1\}}$, then

$$m_n(x) = \frac{\sum_{i=1}^n I_{\{\|x - X_i\| \leq h\}} Y_i}{\sum_{i=1}^n I_{\{\|x - X_i\| \leq h\}}},$$

i.e., one estimates $m(x)$ by averaging Y_i's such that the distance between X_i and x is not greater than h.

Our final example of local-averaging estimates is the *k-nearest-neighbor* (*k-NN*) *estimate*. Here one determines the k nearest X_i's to x in terms of distance $\|x - X_i\|$ and estimates $m(x)$ by the average of the corresponding Y_i's. More precisely, for $x \in \mathbb{R}$, let

$$(X_{(1)}(x), Y_{(1)}(x)), \ldots, (X_{(n)}(x), Y_{(n)}(x))$$

be a permutation of

$$(X_1, Y_1), \ldots, (X_n, Y_n)$$

such that

$$|x - X_{(1)}(x)| \leq \cdots \leq |x - X_{(n)}(x)|.$$

The *k*-NN estimate is defined by

$$m_n(x) = \frac{1}{k} \sum_{i=1}^k Y_{(i)}(x).$$

If $k = k_n \to \infty$ such that $k_n/n \to 0$, then the *k*-NN regression estimate is consistent.

Least-squares estimates are defined by minimizing the empirical L_2 risk

$$\frac{1}{n} \sum_{i=1}^n |f(X_i) - Y_i|^2$$

over a general set of functions $\mathcal{F}_n$. Observe that it does not make sense to minimize the empirical L_2 risk over all functions f, because this may lead to a function which interpolates the data and, hence, is not a reasonable estimate. Thus, one has to restrict the set of functions over which one minimizes the empirical L_2 risk. Examples of possible choices of the set $\mathcal{F}_n$ are sets of piecewise polynomials with respect to a partition $\mathcal{P}_n$, or sets of smooth piecewise polynomials (splines). The use of spline spaces ensures that the estimate is a smooth function. An important member of least-squares estimates is the set of generalized linear estimates. Let $\{\phi_j\}_{j=1}^{\infty}$ be real-valued functions defined on $\mathbb{R}$ and let $\mathcal{F}_n$ be defined by

$$\mathcal{F}_n = \left\{ f; \, f = \sum_{j=1}^{\ell_n} c_j \phi_j \right\},$$

where c_j is the coefficient and ℓ_n is the number of the coefficients. Then, the generalized-linear estimate is defined by

$$m_n(\cdot) = \underset{f \in \mathcal{F}_n}{\arg\min} \left\{ \frac{1}{n} \sum_{i=1}^{n} (f(X_i) - Y_i)^2 \right\}$$

$$= \underset{c_1, \dots, c_{\ell_n}}{\arg\min} \left\{ \frac{1}{n} \sum_{i=1}^{n} \left(\sum_{j=1}^{\ell_n} c_j \phi_j(X_i) - Y_i \right)^2 \right\}.$$

If the set

$$\left\{ \sum_{j=1}^{\ell} c_j \phi_j; \; (c_1, \dots, c_\ell), \, \ell = 1, 2, \dots \right\}$$

is dense in the set of continuous functions, $\ell_n \to \infty$ and $\ell_n/n \to 0$, then the generalized linear-regression estimate defined above is consistent. For least-squares estimates, other examples are neural networks, radial basis functions, orthogonal series estimates or splines.

6.4. General Case: Pricing for Process with Stationary Multiplicative Increments

6.4.1. *The Backward-Recursive Estimation Scheme*

Using the recursion (6.3), if the function $q_{t+1}(x)$ were known, then $q_t(x)$ would be a regression function, which can be estimated from data

$$\mathcal{D}_t = \{(X_{i,t}, Y_{i,t})\}_{i=1}^{n},$$

with

$$Y_{i,t} = \max\{f_{t+1}(X_{i,t+1}), q_{t+1}(X_{i,t+1})\}.$$

However, the function $q_{t+1}(x)$ is unknown. Once we have an estimate $q_{t+1,n}$ of q_{t+1}, we can obtain an estimate of the next q_t by generating samples $\mathcal{D}_t$ with

$$Y_{i,t}^{(n)} = \max\{f_{t+1}(X_{i,t+1}), q_{t+1,n}(X_{i,t+1})\}.$$

6.4.2. *The Longstaff–Schwartz (LS) method*

In this section, we briefly survey recent papers which generalized or improved the Markov chain Monte Carlo and/or LS methods.

First, we recall the original method developed by [Longstaff and Schwartz (2001)], then we elaborate on some refinements and variations. All these methods have the following basic characteristics. They assume that the price process of the underlying asset is well described by a theoretical model, either the Black–Scholes model or a Markov chain model. In both cases, it is also assumed that we have, from historical data, a perfect estimate of the model parameters. Therefore, Monte Carlo generation of an arbitrarily-large number of sample paths of the price process provides arbitrarily-good approximation of the real situation, i.e., one applies Monte Carlo sampling.

[Longstaff and Schwartz (2001)] suggested a quadratic regression as follows. Given that q_t is expressed by a conditional expectation (6.2), we can seek a regression function which determines the value of q_t. Let us consider a function space L_2 and an orthonormal basis, the weighted Laguerre polynomials

$$L_0(x) = \exp(-x/2)$$
$$L_1(x) = (1 - x) L_0(x)$$
$$L_2(x) = \left(1 - 2x + x^2/2\right) L_0(x)$$
$$\vdots$$
$$L_k(x) = \frac{e^x}{k!} \frac{d^k}{dx^k} \left(x^k e^{-x}\right).$$

We determine the coefficients in the case of $k = 2$,

$$(a_{0,t}, a_{1,t}, a_{2,t})$$
$$= \underset{(a_0, a_1, a_2)}{\arg\min} \sum_{i=1}^{n} \left(a_0 L_0(X_{i,t}) + a_1 L_1(X_{i,t}) + a_2 L_2(X_{i,t}) - Y_{i,t}\right)^2$$

and we obtain the estimate of q_t

$$q_{t,n}(x) = \sum_{i=0}^{2} a_{i,t} L_i(x).$$

Other choices might be Hermite, Legendre, Chebyshev, Gegenbauer, Jacobi, trigonometric or even power functions.

[Egloff (2005)] suggested replacing the parametric regression in the LS method by nonparametric estimates. For example, using the generated variables, one can obtain the least-square estimate of q_t by

$$q_{t,n} = \arg\min_{f \in \mathcal{F}} \left\{ \frac{1}{n} \sum_{i=1}^{n} (f(X_{i,t}) - Y_{i,t})^2 \right\},$$

where $\mathcal{F}$ is a function space.

[Kohler (2008)] studied the possible refinement and improvement of the LS method in several papers. One significant extension is the computational adaptation of the original LS method to options based on d underlying assets. To analyze this d-dimensional time-series [Kohler (2008)] suggested a penalized spline estimate over a Sobolev space.

[Kohler *et al.* (2010)] investigated a least-squares method for empirical-pricing of compound American options if the corresponding space of functions $\mathcal{F}$ is defined by neural networks.

[Egloff *et al.* (2007)] reduced the error propagation with the rule that the non-*in the money* paths are sorted out and for $(X_{i,s}, Y_{i,s})$ generate new path working on $t, \ldots, T$ (instead of $t+1, \ldots, T$). They studied an empirical error-minimization estimate for a function space of polynomial splines.

6.4.3. *A New Estimator*

Let us introduce a partitioning-like estimate, i.e., for the grid $\mathbf{G}$ and for $x \in \mathbf{G}$ put

$$q_{t,n}(x) = \frac{\sum_{i=1}^{n} \max \left\{ f_{t+1}(X_{i,t+1}), q_{t+1,n}(X_{i,t+1}) \right\} I_{\{|X_{i,t}-x| \leq h/2\}}}{\sum_{i=1}^{n} I_{\{|X_{i,t}-x| \leq h/2\}}},$$

where I denotes the indicator, and $0/0 = 0$ by definition. Obviously, this estimate should be slightly modified if the denominator of the estimate is not large enough. Then, linearly interpolate for $x \notin \mathbf{G}$.

We have that

$$\max_{s \in \{t+1,\ldots,T\}} \mathbb{E}\left\{ f_s(X_s) \,|\, X_t = x \right\} \leq q_t(x) \leq \mathbb{E}\left\{ \max_{s \in \{t+1,\ldots,T\}} f_s(X_s) \,|\, X_t = x \right\},$$

where both the lower and upper bounds are true regression functions.

For $x \in \mathbf{G}$, the lower bound can be estimated by

$$q_{t,n}^{(l)}(x) = \max_{s \in \{t+1,\ldots,T\}} \frac{\sum_{i=1}^{n} f_s(X_{i,s}) I_{\{|X_{i,t}-x| \leq h/2\}}}{\sum_{i=1}^{n} I_{\{|X_{i,t}-x| \leq h/2\}}},$$

while an estimate of the upper bound is

$$q_{t,n}^{(u)}(x) = \frac{\sum_{i=1}^{n} \max_{s \in \{t+1,\dots,T\}} f_s(X_{i,s}) I_{\{|X_{i,t}-x| \leq h/2\}}}{\sum_{i=1}^{n} I_{\{|X_{i,t}-x| \leq h/2\}}}.$$

Again, a truncation is proposed:

$$\hat{q}_{t,n}(x) = \begin{cases} q_{t,n}^{(u)}(x) & \text{if} \quad q_{t,n}^{(u)}(x) < q_{t,n}(x), \\ q_{t,n}(x) & \text{if} \quad q_{t,n}^{(u)}(x) \geq q_{t,n}(x) \geq q_{t,n}^{(l)}(x), \\ q_{t,n}^{(l)}(x) & \text{if} \quad q_{t,n}(x) < q_{t,n}^{(l)}(x). \end{cases}$$

References

Carrier, J. (1996). Valuation of early-exercise price of options using simulations and nonparametric regression, *Insurance: Mathematics and Economics* **19**, pp. 19–30.

Chow, Y. S., Robbins, H. and Siegmund, D. (1971). *Great Expectations: The Theory of Optimal Stopping* (Houghton Mifflin, Boston).

Egloff, D. (2005). Monte Carlo algorithms for optimal stopping and statistical learning, *Annals of Applied Probability* **15**, pp. 1–37.

Egloff, D., Kohler, M. and Todorovic, N. (2007). A dynamic look-ahead Monte Carlo algorithm for pricing American options, *Annals of Applied Probability* **17**, pp. 1138–1171.

Gelencsér, G. and Ottucsák, G. (2006). NYSE data sets at the log-optimal portfolio homepage, www.szit.bme.hu/~oti/portfolio.

Glasserman, P. (2004). *Monte Carlo Methods in Financial Engineering* (Springer, New York).

Györfi, L., Kohler, M., Krzyżak, A. and Walk, H. (2002). *A Distribution-Free Theory of Nonparametric Regression* (Springer, New York).

Haugh, M. B. and Kogan, L. (2004). Pricing American options: a duality approach, *Operation Research* **52**, pp. 258–270.

Kohler, M. (2008). A regression based smoothing spline Monte Carlo algorithm for pricing American options, *Advances in Statistical Analysis* **92**, pp. 153–178.

Kohler, M. (2010). Review on regression-based Monte Carlo methods for pricing American options, in L. Devroye, B. Karasözen, M. Kohler and R. Korn (eds.), *Recent Developments in Applied Probability and Statistics, dedicated to the Memory of Jürgen Lehn* (Physica-Verlag Heidelberg, Berlin), pp. 37–58.

Kohler, M., Krzyżak, A. and Todorovic, N. (2010). Pricing of high-dimensional American options by neural networks, *Mathematical Finance* **20**, pp. 383–410.

Kohler, M., Krzyżak, A. and Walk., H. (2008). Upper bounds for bermudan options on Markovian data using nonparametric regression and areduced number of nested Monte Carlo steps, *Statistics and Decision* **26**, pp. 275–288.

Longstaff, F. A. and Schwartz, E. S. (2001). Valuing American options by simulation: a simple least-squares approach, *Review of Financial Studies* **14**, pp. 113–147.

Luenberger, D. G. (1998). *Investment Science* (Oxford University Press, Oxford).

Rogers, L. C. G. (2002). Monte Carlo valuation of American options, *Mathematical Finance* **12**, pp. 271–286.

Shiryaev, A. N. (1978). *Optimal Stopping Rules* (Springer-Verlag, Berlin).

Stout, W. F. (1974). *Almost sure convergence* (Academic Press, New York).

Tsitsiklis, J. N. and Roy, B. V. (1999). Optimal stopping of Markov processes: Hilbert space theory, approximation algorithms, and an application to pricing high-dimensional financial derivatives, *IEEE Transactions on Autom. Control* **44**, pp. 1840–1851.

Tsitsiklis, J. N. and Roy, B. V. (2001). Regression methods for pricing complex American-style options, *IEEE Transactions on Neural Networks* **12**, pp. 694–730.

Vajda, I. (2006). Analysis of semi-log-optimal investment strategies, in M. Huskova and M. Janzura (eds.), *Prague Stochastics 2006* (MATFYZ-PRESS, Prague).

Index